The Book of Instructions in the Elements of the Art of Astrology

by Al Biruni

The Book of Instruction

In the Elements of the Art of

ASTROLOGY

By

Abu'l-Rayhan Muhammad Ibn Ahmad

AL-BIRUNI

Written in Ghaznah, 1029 A.D.

The Translation by
R. Ramsay Wright, M.A. Edin., LL.D. Tor. and Edin.
Emeritus Professor of Biology
University of Toronto

1934
LONDON
LUZAC & CO.

2006
Astrology Classics

Thanks to Rob Hand for the scans.

ISBN 1 933303 16 6

In preparing this edition I have deleted footnotes concerning differences of opinion regarding text or translation, as well as notes
not related to astrological matters. I have also added the index.
David R. Roell, Publisher.

Published by
Astrology Classics

The publication division of
The Astrology Center of America
207 Victory Lane, Bel Air MD 21014

On the net at www.**AstroAmerica.com**

Table of Contents

The paragraphs are referred to in the text by numbers alone.

DIVISIONS OF THE SIGNS.

THE HOUSES.

THE PART OF FORTUNE.

JUDICIAL ASTROLOGY.

ASTROLOGY

347. NATURE OF THE SIGNS. And first we shall deal with the relation of the signs to the characteristics of the four elements, separately and in combination.

As to the nature and temperament of the signs if they are written down in two rows, upper and lower, the first sign above and the second below it, and so on to the last, all those of the upper row are hot and those of the lower cold, while the pairs so arranged are alternately dry and moist.

	Dry	Moist	Dry	Moist	Dry	Moist
Hot	Aries	Gemini	Leo	Libra	Sagittarius	Aquarius
Cold	Taurus	Cancer	Virgo	Scorpius	Capricornus	Pisces

When therefore you know the active virtues of a sign whether heat or cold, and the passive virtues, whether dryness or moisture, it will not be concealed from you what particular element of the world and what particular humour of the body each sign resembles. Each sign that is hot and dry is related to fire and yellow bile, each that is cold and dry, to earth and black bile, each that is hot and moist to air and blood and each that is cold and moist to water and phlegm.

The Hindus regard as moist signs Pisces, the hinder half of Capricorn and the anterior half of Aquarius for reasons given above in speaking of their representations, viz. that the hinder end of Capricorn is fish-like, and that of Aquarius water. They do not however reckon Scorpius as belonging to the moist signs, but count it with the aerial ones, while Cancer holds an intermediate position, sometimes being regarded as watery, sometimes as aerial according to circumstances.

348. MALE AND FEMALE. All the hot signs are male and the cold female. The planets are powerful in those signs which resemble

7

them in nature and sex, but they partake of the nature of the signs in which they are situated so that a planet obviously male shows a tendency to femaleness by being in a female sign. The Hindus say that all the odd, i.e. male signs are unlucky and the female signs lucky.

349. DIURNAL AND NOCTURNAL. There is a general agreement that all the male sign are diurnal and the female nocturnal. The diurnal planets are powerful in the day signs and the nocturnal in the night ones. In the Greek bizidhaj it is stated that according to some Aries, Cancer, Leo and Sagittarius are day signs and their nadirs Libra, Capricorn, Aquarius and Gemini are night ones, while the remainder partake both of day and night. The Hindus believe that Aries, Taurus, Gemini, Cancer, Sagittarius and Capricorn are powerful at night, the six others by day.

350. MAIMED. Aries, Taurus, Leo and Pisces are described as maimed, the first three because their feet are cut off at the hoofs and claws, and Taurus in addition because it is only half an ox cut in two at the navel, while Pisces is included on account of the absence of limbs.

351. ERECT AND OTHERWISE. Aries, Libra and Sagittarius are described as erect constellations in the books, the others are not referred to in this regard, but the Hindus always say that Aries, Taurus, Cancer, Sagittarius and Capricornus are asleep and represent them recumbent, while Leo, Virgo, Libra, Scorpius and Aquarius are erect, and Gemini and Pisces recline on one side. Their intention in this matter is unknown to me, for the position of the figures in the constellations is of no importance, and they offer no evidence to the contrary.

352. HUMAN AND OTHERWISE. The following signs are represented as human: Gemini, Virgo, Libra and half of Sagittarius and Aquarius. Such is the case in the figures shown above with the exception of Libra, but when Libra is represented in the act of weighing, a human or bird figure suspends the balance or simply a human hand. The four-footed figures are Aries, Taurus, and Leo, while the hinder half of Sagittarius, sometimes the front half (of Capricorn on the analogy of Taurus) are also so reckoned. Then of these Aries and Taurus have cloven feet, Leo claws and Sagittarius hoofs. Again the people generally from youth up entertain certain ideas as to the signs, such as that Leo, Scorpius, Sagittarius and Capricorn (Pisces) suggest beasts of prey; Gemini, Virgo, Pisces and the hinder two-thirds of Capricorn, birds; Cancer, Sagittarius, Scorpius and Capricorn, reptiles; and Cancer, Scorpius and Pisces, aquatic animals.

The Hindus have a redundancy of interpretations of this kind; they say that the human signs are Gemini, Virgo, Libra (the fore part of Sagittarius) and the hinder half of Aquarius, all of which they describe as bipeds, while the quadrupeds are Aries, Leo, the hinder half of Sagittarius, and the fore part of Capricorn. Reference has already been made to their ideas as to watery and aerial signs.

353. VOICED AND VOICELESS. Gemini, Virgo and Libra are loud-voiced, of these Gemini is capable of speech; Aries, Taurus and Leo are half-voiced, Capricorn and Aquarius are weak-voiced, while Cancer, Scorpius and Pisces are voiceless. Knowledge as to voice and speech is essential as to whether in a difficulty indications in these signs are harmful or the reverse.

354. FERTILE AND BARREN. Indications of the signs as to families. The watery signs Cancer, Scorpius, Pisces and the hinder half of Capricorn favour large families; Aries, Taurus, Libra, Sagittarius and Aquarius small ones, while the first part of Taurus, Leo, Virgo and the first part of Capricorn indicate sterility. The production of twins is specially in charge of Gemini, but also is favoured by Virgo, Sagittarius and Pisces, and sometimes by Aries and Libra and the last part of Capricorn. (The fore part of Capricorn and Scorpius indicate hermaphroditism.) In consequence of what we have said Aries and Libra are described as being of two natures, as are also Capricorn and Sagittarius. Virgo is called mistress of three forms, and Gemini as many-faced, because they denote not only twins but three or more children.

355. RELATION TO MARRIAGE. With regard to marriage, Aries, Taurus, Leo, Capricorn indicate eagerness therefor, for Libra and Sagittarius much the same can be said. With regard to the conduct of women, Taurus, Leo, Scorpius and Aquarius denote reserve and abstinence; Aries, Cancer, Libra and Capricorn corruption and bad conduct, while Gemini, Virgo, Sagittarius and Pisces denote a mean in this regard; of the four Virgo is the most virtuous.

356. DARK AND ANXIOUS SIGNS. Leo, Scorpius and Capricorn are dark and anxious, and there is a suspicion of trouble in Virgo and Libra.

357. RELATION TO POINTS OF COMPASS. Aries denotes the middle of the East, Leo a point to the left of that towards the North, and Sagittarius one to the right towards the South; similarly with each of the other triplicities. Thus Taurus indicates the centre of the South, Virgo a point to its left towards the East and Capricorn one

9

to the right and West. Gemini occupies the centre of the West, Libra a point to its left and South and Aquarius, one to its right and North, Cancer is in the centre of the North, Scorpius a point to the left and West and Pisces to the right and East. All are represented in the accompanying figure.

358. RELATION TO WINDS. A wind coming from a quarter associated with a particular sign is also associated with that sign thus the

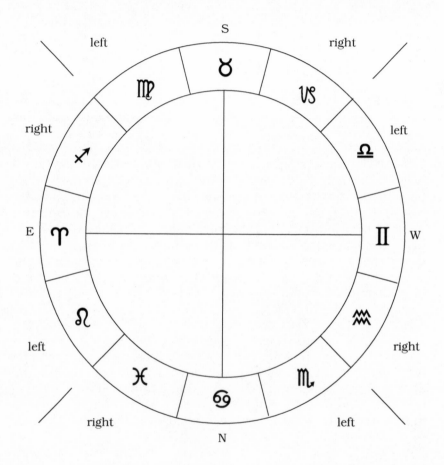

East wind with Aries, the West with Gemini, the South with Taurus and the North with Cancer. Similarly with the intermediate quarters, a S.E. wind is related to Virgo or to Sagittarius according as it is nearer S. or E.

359. RELATION TO PARTS OF BODY. The following are the various parts of the body which are related to the several signs. The head and face to Aries, the neck and windpipe to Taurus, the arms and hands to Gemini, the chest, breasts, sides, stomach and lungs to Cancer, the heart to Leo, the womb with its contents to Virgo, the back and buttocks to Libra, the genitals to Scorpius, the thighs to Sagittarius, the knees to Capricorn, the shanks to Aquarius and the feet and heels to Pisces.

In this matter there is much confusion in the books, for according to some, not only the head and face but also the bowels are governed by Aries. The analogy in this case does not seem to be clear. But it is clear in the saying of a Brahman that if we imagine the zodiac to be a man, with Aries the head and the soles of the feet directed towards it, then the allocation of the parts of the body according to the Hindus conforms with what we have said above except that the face is given to Taurus.

The signs are also indicative of the various diseases of man, of his complexion, figure, face, and the like, they also govern localities and countries, and denote various matters regarding animals, fire, water, etc. To facilitate study these are set down in the accompanying tables.

But God is All-knowing.

♈	Laughing and talkative, kingly and haughty, fond of poetry, sharp-tongued, lustful, brave.
♉	Of good judgment, negligent, a liar, a cheat, lustful and a fool.
♊	Generous, chaste, excelling in games, fond of philosophy and astronomy, munificent, violent, and a hafiz (has the Qur'an by heart).
♋	Indolent, dumb, fickle and changeable.
♌	Kingly, formidable, sharp-tongued, hard-hearted, litigious, knavish, many troubles, a sinner, forgetful, powerful by nature, bold.
♍	Liberal, good manners, truthful, well-informed, pious, a judge, thoughtful, lively, playful, fond of dance and music, a hafiz.
♎	Thoughtful, polite, generous, indolent, cowardly, just judge, plebeian, excited in speech, a musician and singer, a hafiz.
♏	Generous by nature, anxious, deceitful, bold, rough morose, sharp-tongued, a slayer, a hafiz, a fool, indolent, pleased with himself, bold.
♐	Kingly, reticent, liberal, tricky, prejudiced, a capable mathematician, surveyor, thoughtful about the next world, fond of horses, particular as to food, drink and clothing, virile.
♑	Arrogant, false, choleric, impetuous, changeable, evil-thinking, anxious, quarrelsome, opinionative fond of games and life, crafty, forgetful, shaqq, bold.
♒	Well-disposed chaste, eager to accumulate riches, eager for magnificence and manliness, a gourmet, bad-hearted, inert, indolent, restful, too anxious about worldly affairs.
♓	Good disposition, generous, elegant, lustful, unstable in his opinions, of good faith, mediocre in business, tricky and deceitful, liable to err, forgetful, foolish, bold.

♈	Medium height, thin, shortsighted, glance upcast, eyes dark, or gray, or dark gray, nose and ears large, ugly mouth, hair curly and reddish.
♉	Tall, broad forehead, eyebrows short, eyes black, the whites small, downcast, nose broad, the point upturned, large mouth, thick lips, hair black, neck strong.
♊	Medium height, good appearance, erect, fine beard and face, sharp-sighted, broad-shouldered, shanks long in comparison with forearm.
♋	Moderate height, limbs thick rather long, hair fine brown, nose crooked, teeth uneven, downcast look, corpulent, shanks longer than forearms.
♌	Full height, broad face, thick fingers, slender thighs, hip bigger, good-looking, gray-eyed, accustomed to wine, large nose, wide mouth, (teeth separated from each other), chestnut hair, prominent belly.
♍	Medium stout inclining to tall, long hair, moles on chest and abdomen, broad shoulders and chest, flat-nosed.
♎	Moderate size, good-looking, colour inclined to brown and yellow, gray eyes, good nose, distinctive marks on neck and waist, good feet.
♏	Head erect, good-looking, eyes small, whites yellow, face round, forehead narrow, hair coarse, slender thighs and ankles, broad chest and shoulders, broad nose, paunch and a mark on the back.
♐	Light, tall, good-looking especially from back front view, good eyes, long beard, coarse nose, complexion red, belly large, shanks longer than thighs, marks on arms and legs.
♑	Body slender, erect, fine figure, face goat-like, wide gray eyes, ears crooked, long beard, little hair, thin legs, active gait, handsome.
♒	Medium height, tending to tall, forehead narrow, eyes dark gray, black, the black part wider than the white, coarse lips, downcast look, body well filled out, legs unequal, good-looking, broad chest.
♓	Good figure, delicate joints, smooth skin, fine face, medium stature, fairly broad chest, narrow shoulders, small head, narrow forehead, looks down, black eyes, handsome.

	362-364 COLORS (Hindu opinion), CLASSES AND ARTISANS
♈	White and reddish. Reddish white. Kings, bankers, coiners, blacksmiths, coppersmiths, butchers, shepherds, spies and thieves.
♉	White and brownish not shining. White. Sellers (Tailors), and weighers of grain, fishermen, (cobblers), agents and farmers.
♊	Greenish yellow. Pistachio green. Kings, calculators, teachers, hunters, dancers, musicians, painters, tailors.
♋	Smoke-coloured not quite black. Dark red. Sailors, (water diviners, swimmers) and canal-diggers.
♌	Whitish red. White of clothes, withered vegetables. Horsemen, coiners, falconers.
♍	Whitish yellow. Changing. Vazirs, eunuchs, secretaries, supervisors, ordinary people, dancers, singers, assemblies of men.
♎	White tinged black. Back. Magnates and dignitaries, privy counsellors, merry-makers, philosophers, geometricians, merchants, (grammarians), devotees.
♏	(Vacant P.B. and M.) Golden. Physicians, enchanters, magicians, sailors.
♐	Reddish. Colour of palm fibres. Horse-dealers, middle-class people, busybodies, meddlers with other people's business, (who, although wit honest intentions, excite strife), undertake their burdens.
♑	Colours mixed like a peacock, brown and green. Piebald, black and white. Hunters and slaves.
♒	Yellow, sapphire blue and various colours. Bright red turning yellow. Servants, traders, ass-drivers, makers of glass and jewelry, uneducated people, grave-robbers.
♓	White. Khaki. Most revered and religious people. The last part of the sign for blind men, those who operate on them for cataract and sailors.

♈	Babylon, Fars, Palestine, Adharbaijan, Alan.
♉	Districts of Iraq, Mahin, Hamadhan, Mountains of Kurdistan, Ctesiphon, Cyprus, Alexandria, Constantinople, Oman, Rai, Farghana, and shares in the control of Herat and Sijistan.
♊	Egypt, the cities of Barqa, Armenia, Gurgan, Gilan, Muqan and shares in Isfahan and Kirman.
♋	That part of Armenia Minor which is beyond Muqan. Parts of Africa, Hajar, Bahrain, Dabil, Marwalrudh, Eastern Khurasan, and shares in Balkh and Adharbadgan.
♌	Turkestan as far as Gog and Magog, and the ruined cities there, Ascalon, Jerusalem, Nisibis, the twin cities, Malatya, Sistan, Makran, Dailam, Abrashahr, Tus, Soghdiana, Tirmidh.
♍	Andalusia, Syria, Crete, the Euphrates and Mesopotamia, Jararamaqa, the capital of Abyssinia, San'a, Kufa, the cities of Fars in the direction of Kirman, and Sistan as far as the borders of India.
♎	The Greek Empire as far as Tunisia, and upper Egypt to the confines of Abyssinia, Antioch, Tarsus, Mecca, Taliqan, Tokharistan, Balkh, Herat, Sistan, Kabul, Kashmir and China.
♏	The Hijaz country, the desert of Arabia as far as Yemen, Tangier, Qiyad, Khazaria, Qumis, Amul, Sariah, Nahawand, Nahrawan, and shares in Turkish Soghdia.
♐	Persian Iraq, Dinawar, Isfahan, Rai, Baghdad, Danbavand, Darband of the Khazars, Jundi-Sabur, shares in Bukhara and Gurgan, the borders of the Sea of Armenia and Barbary as far as Morocco.
♑	Makran and Sind, and the river Mihran (Indus) and the sea between Oman and Hindustan, Eastern China, Asia Minor, Ahwaz and Istakhr (Persepolis).
♒	Southern Iraq as far as Kufa and Hijaz, the country of the Copts, the West of Sind and shares in Fars.
♓	Tabaristan and the country north of Gurgan, Bukhara and Samarqand, shares control in Asia Minor: Qaliqala as far as Syria, Mesopotamia, Egypt, Alexandria, the sea of Yemen and Eastern Hindustan.

	366 AS TO PLACES.
♈	Deserts, pasturing places for beasts of burden. Woodsheds, places where fire is used, thieves' dens, places where jewelry is manufactured.
♉	Mountainous places, orchards, pasture land, storehouses for food, cow and elephant sheds.
♊	Mountains, hills, mounds, hunting-grounds, riversides, resorts of acrobats and gamblers and musicians, kings' palaces.
♋	Reservoirs, reed-beds, river margins, cultivated places, trees, wells, rivers, and places of worship.
♌	Mountains, fortresses, high sanctuaries, kings' palaces, desert places, quarries, barren saltish ground.
♍	Divans, women's quarters, musicians' houses, threshing floors cultivated fields.
♎	Small mosques and places of worship, castles, cultivation, palm-groves, observatories, plains, orchards, tops of mountains which are cultivated.
♏	High places, pools of bad water, prisons, places of grief and mourning, scorpions' holes, deserted places, vineyards, mulberry-groves.
♐	Level plains, Magian temples, Christian churches, arsenals, cattle-stalls, lime-pits, irrigated orchards.
♑	Castles, ancient reservoirs, harbours, fireplaces, (weeping places), slaves' sleeping places, holes of dogs and foxes, lodgings for strangers. The first part of the sign indicates stone and gravel and water wheels.
♒	Running and standing water, heated bath-water, taverns, brothels, canals and ditches, birds nests and resorts of aquatic birds.
♓	Abodes of angels, holy men, Magian priests, mourning places, canebrakes, lake shores, salt marshes, granaries.

16

	367-369 TREES & CROPS; WATER, WIND & FIRE; JEWELS & FURNITURE.
♈	——- . Fire is used. Copper, iron, lead; helmets, diadems, crowns and girdles.
♉	Unirrigated fields, crops from setting out cuttings. ——. Clothes, necklaces, wool, hair, collars; sweet fruits, artichokes, bastard saffron.
♊	Tall trees. Zephyr, gentle winds, animal spirits. Armlets, bracelets, dirams, dinars, attar; drums lutes and flutes.
♋	Tall and medium trees. Good drinking water, rain, running water, and that which comes down from the sky. Rice and cane sugar.
♌	Tall trees. Torrents, subterranean fires, minerals extracted from the ground, cloudy weather. Coats of mail and cuirasses, tall metal vessels; emeralds and rubies, gold and silver objects manufactured from them.
♍	Sown fields, sowing and planting. All running water. Mercury; (berries, herbs and the ordinary seeds.)
♎	Date palms, tall trees, and such as are grown on the top of mountains. Winds which favour trees and fruits, which make trees large and spread them; denotes dark atmosphere. Silks, lutes and drums.
♏	Medium sized trees. Running waters, rivers, torrents, underground conduits, black mud and drowned land, such articles as are kneaded of clay. Precious stones from water, like coral; (drugs), sal-ammoniak, water vessels, awani, such things as are made with fire.
♐	——— . Natural streams and heat in the bodies of animals. Tin, gold, all manufactured articles, arrows and (bows and) spears and armour, (earthenware) garments, armour, nibs (harf) (burnt brick and lime depilatory.
♑	Crops, herbage and the like, such as do not require to be sown, fruit. ——— . ——— .
♒	Tall trees, plantain and ebony, myrobalan and belleric myrobalan. Seas, running waters, winds which stir up the seas, and destroy tall trees and herbage; cold fogs. Tools and sites for drawing water and for building houses, and for digging and planting trees.
♓	Cotton, sugar, fruit-bearing trees, sandal wood, camphor, edible fruits. First half, medium-sized trees. Stagnant waters & lakes. Pearls, mother of pearl, coral; shoes, clogs, soles.

	370. SICKNESS AND DISEASE
♈	At first very strong, afterwards weak and liable to disorders, especially in the head such as baldness, blood to the face, rashes, lepra and scab, limbs worn out, phlegmatic, sweet-smelling.
♉	At first very strong, towards the end lean and spare, only moderately subject to disorders, for the most part of the neck like scrofula, and quinsy and points to freckles, ozaena and marks on back and breast.
♊	Healthy and sweet-smelling body, illnesses not serious, generally catarrh or gout, not much distress.
♋	Weak and sickly, gout, catarrh, cancer ,baldness, eczema, deafness, ringworm, dandruff, leprosy, pimples, piles, heaviness in left foot and fingers.
♌	At first strong, but afterwards weak and liable to disease, especially of the stomach and pain in the eyes, loss of hair; at first offensive breath.
♍	Strong, moderately lean, and slender, sickness moderate, loss of hair, sal.
♎	Limbs strong, sound, middling slender.
♏	At first strong and thickset but at the end of life weak and sickly, illnesses chiefly deafness and dumbness, cataract, cancer, eczema, ringworm, leprosy, retention of urine, eunuchism.
♐	At first strong, at last weak and sickly, moderately thin healthy body, gout, catarrh, blindness, blind of one eye, baldness, epilepsy, superfluous fingers, headache, and marks on the legs.
♑	Weak sickly but sound limbs, deaf and dumb, opthalmia, bleeding, itch, scrofula, cancer, baldness, tumours; the tendency to baldness much stronger than under other signs.
♒	At first strong, at last weak and sickly, limbs sound, diseases of the tongue, jaundice, catarrh, gout, bilious headache, pain in the eyes, and veins, vertigo, rupture, epilepsy and ozaena.
♓	Weak, thin, sickly, especially in limbs, (nerves a'sab) gout, sleeping of the limbs, bilious, eczema, ringworm, dandruff, bald, sal, leprosy, catarrh and abundant hair, athith.

	371. AS TO VARIOUS ANIMALS
♈	All hoofed animals, wild and domestic such as goats and sheep; also rams and deer.
♉	Cows, calves, elephants, gazelles; animals which become attached to man.
♊	Domestic fowls and such birds as become tame; gazelles and horned vipers.
♋	Reptiles, aquatic and terrestrial animals, that are numerous in the desert like beetles; poisonous lizards.
♌	Wild horses, tame lions, all animals with claws, black snakes
♍	Magpies, black crows, bulbuls, sparrows, parrots, large serpents.
♎	Birds, leopards, and jinn.
♏	Reptiles, aquatic animals, destructive wild beasts (of prey), many-footed animals like scorpions and wasps (and poisonous insects).
♐	Solid-hoofed animals especially pack-horses, mules, asses. There is also an indication of birds and reptiles.
♑	Kids, lambs, animals that are herded, creeping things, apes, locusts.
♒	Bipeds, vultures, sinur for nusur, eagles, beavers, jerboas, sinjab, sables, ermines, aquatic birds especially black ones.
♓	Birds, fish, large and small, aquatic carnivores, serpents, scorpions.

372. ON THE YEARS OF THE SIGNS*

	Years	Months	Days	*also* Days	Hours
♈	15	15	37½	4 (3)	3
♉	8	3 (8)	20	1	16
♊	20	20	50	4	4
♋	25	25	62½	5	5
♌	19	19	47½	3	23
♍	20	20	50	4	4
♎	8	8	20	1	16
♏	15	15	27½	3	3
♐	12	30	30	2	12
♑	27	27	67½	5	15
♒	30	30	75	6	6
♓	12	2 (12)	30	2	(12) 2

373. IN ASPECT OR INCONJUNCT. As the complex must follow the simple we have now to consider the relations of the signs to each other.

Each sign is in sextile aspect to the third and eleventh left and right of it, and there is a sixth of the zodiac (60°) between any degree of that sign and the same degree of those named. Similarly the quartile aspect is between a sign and the fourth and tenth left and right, separated by 90°, and the trine between the fifth and ninth, distant 120° and the opposite sign is the seventh, 180°. There are therefore seven signs to which the sign in question turns its face and which are consequently considered to be bound in aspect to it. The two signs which are each side of the one in question and their opposites, viz. the second and twelfth and the sixth and eighth are not in aspect and are known as inconjunct.

---

* No explanation is given of this table. It is arrived at by the second of two methods described in Abu Ma'shar's Madkhal f. 235. The years and the months are equal in number to the minor years of the lord of each sign and the days and hours are the same number multiplied either by 5/2 or by 5/24.

No reason is given for the two domiciles of Saturn being allotted different numbers. (They are the same (30) in the Opus Introd. Venice, 1506 where the four last columns are unexplained.) But Vettius Valens p.164, gives a reason; he assigns 1/4 of the Sun's great years to Aquarius, and 1/4 of the Moon's to Capricorn [two Greek words].

The first method allows a year for every degree of oblique ascension of each sign in any climate and a month for every five minutes. The numbers in brackets are the correct ones.

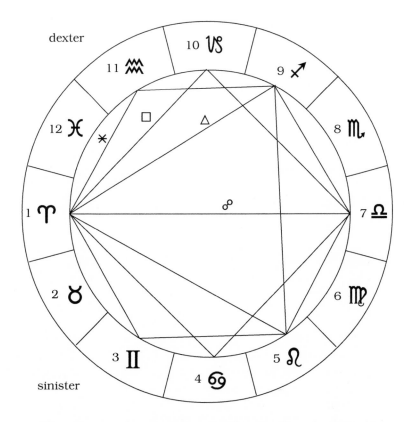

The signs used are conjunction ♂, opposition ☍, sextile ✶, quartile ☐, trine △.

374. SIGNS FRIENDLY, UNFRIENDLY, HOSTILE. Signs which are in sextile or trine are friendly to each other, those in quartile unfriendly and that opposite inimical. Thus Gemini and Aquarius are in sextile to Aries, Leo and Sagittarius in trine to it, and these are mutually friendly, while Cancer and Capricorn being in quartile and Libra opposite are inimical. The inconjunct signs to Aries are four, viz. Taurus, Virgo, Scorpius & Pisces.

375. RELATIVE POWER OF ASPECTS. The following is the order in power of the various aspects. The most powerful is conjunction, i.e. meeting in the same sign, then the opposite, then the dexter* quartile, sinister quartile, dexter trine, sinister trine, dexter sextile, sinister sextile. When there are two aspects the more powerful renders the weaker one incompetent and takes away its power.

---

* Dexter aspects are those contrary to the order of the signs, so that a planet in Aries casts a dexter quartile to one in Capricorn and a sinister quartile to one in Cancer.

376. HINDU OPINION. The Hindus are partly in agreement and partly dissent from this doctrine. They are in agreement in regard to the opposite, quartiles and trines, but they say that while a sign looks towards its third, the third does not regard it, and while it does not look to its sixth, the sixth does regard it. They do not apply the term aspect to conjunction, for they say that when one stands erect and looks ahead, one cannot see oneself. With regard to the relative importance of the aspects they say that from a sign towards the third and tenth signs there is a quarter of an aspect, and to the fifth and ninth, half an aspect (towards the 8th and 4th three quarters of an aspect and to the 7th a complete aspect). They describe the second and twelfth signs as inconjunct to the first and it to them.

377. RELATIONS BESIDES ASPECT. Two signs equidistant from an equinoctial point are said to be equipollent, because the day hours of each are equal to the night hours of the other, and their ascensions are equal in all places, such as Aries and Pisces, Taurus and Aquarius, etc. The correspondence is by inverse degrees, one being north the other south, the first of Aries being equal to the twenty-ninth of Pisces,* and the 10th to the 20th.

Two signs revolving in the same parallel, North or South (equidistant from a solstice) are described as corresponding in course (in itinere), their day hours are equal as are their night hours, and their ascensions are identical at the equator, such as Gemini and Cancer, Taurus and Leo. The correspondence is also by inverse degrees, the beginning of Cancer corresponding to the end of Gemini, and the tenth of the former to the twentieth of the latter. These two relations receive different names in the books, and there is no permanence in such names, but that term is best which corresponds to the meaning.

Abu Ma'shar has called the two signs which have the same presiding planet as concordant in itinere, and although this is different from the two kinds of agreement referred to above, it is a relation which has to be considered. With regard to the agreement which we have spoken of, Abu Ma'shar calls the relation of Aries to Pisces, and of Virgo to Libra by power, and that of Gemini to Cancer and Sagittarius to Capricorn by course, as natural sextiles, although they do not regard each other, but since the nearest aspect to the inconjunct place is the sextile, he has called them by that name. Similarly the relations of Aries to Virgo and Pisces to Libra by course, and those of Gemini to Capricorn and Cancer to Sagittarius by power,

* The 30th degree being regarded as destitute of a companion so as to associate odd degrees with odd and even with even.

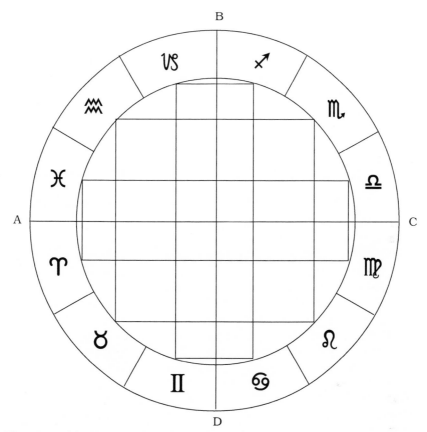

The vertical lines join equipollent signs, the horizontal those corresponding in their course.

ADC: Northern half.   CBA: Southern half.
DAB: Ascending.   BCD: Descending half.

The ascending signs according to Wilson are ♈ ♉ ♊ , ♎ ♏ ♐ because when in them the sun's declination is increasing. *Translator's notes.*

he speaks of as natural opposites, although there is here no aspect. But in the quartile aspect it occasionally happens from these agreements as in the case of Taurus to Aquarius and Leo to Scorpius by power, and in that of Taurus to Leo and Scorpius to Aquarius by course, that the disagreeable enmity of the quartile lessens and its evil influence disappears, so that the significance of the relation gains in power, just as the removal of the inconjunction, obscurity and evil from those sextiles and natural opposites also takes place.

From those two correspondences to which we have adverted, power and course, the zodiac is divided into two sets of halves, 1.

Northern and southern halves, 2. Ascending and descending halves.

378. ASCENDING AND DESCENDING HALVES OF ZODIAC. The latter are marked out by the solstices, the ascending half comprising the following signs: Capricorn, Aquarius, Pisces, Aries, Taurus, Gemini, and the descending half the nadirs of these.

The Hindus call these halves 'ayana', the ascending 'uttarayana' or north, because although the declination of the sun in this half of the ecliptic is south, yet the sun during the whole of the half keeps its face towards its northern goal. The descending half is called 'dakshayana' or southern by similar reasoning.

Signs of the ascending half are described as signs of short or crooked ascension, because their oblique ascension is shorter than that in the erect sphere, while those of the descending half are said to be signs of long or direct ascension, because their oblique ascension is longer than that in the erect sphere.

The crooked signs are also called 'obedient' and this is due to concordance in course, because when you compare two signs on one parallel, the one belonging to the descending half comes first by the diurnal movement, and the one of the ascending half later; so the former commands, the latter, which obeys the command and always follows.*

379. TRIPLICITIES. Those signs whose nature as regards two qualities is identical are situated in the zodiac at the angles of right-angled triangles; they are consequently known as triplicities and arc recognized as entities, although three in number, the effects of each being identical or similar. The first triplicity is formed of Aries, Leo and Sagittarius, all of which are fiery in their nature, withering and heavy, while the special domain of each is for Aries, fires in ordinary use, for Leo those present in minerals and plants, and for Sagittarius that which is distributed from the heart of animals throughout the body.

The second triplicity composed of Taurus, Virgo and Capricorn is earthy, generous with its wealth, and the interpretation of

---

* According to Paulus Alexandrinus, the commanding signs are from Taurus to Virgo. The beholding signs from Gemini to Aquarius. So also Valens p.24. But Chaucer says l.c. p. 38 "These crooked signs [are] obedient to the signs that [are] of right Ascension (Cancer to Sagittarius)." The northern signs command because when the sun is on them, the day is longer than the night. [Edited by the publisher]

24

its effects is that Taurus is responsible for pastureland which is not sown, Virgo for plants which have neither berries nor seeds and small trees, Capricorn for sown crops and large and tall trees.

Gemini, Libra and Aquarius form the third triplicity which is airy in nature, sending winds abroad, and in detail Gemini is characterized by that quiet air which produces and sustains life, Libra by that which causes trees to grow, fertilizes them and produces fruit, and Aquarius by destructive storms.

The fourth triplicity of Cancer, Scorpius and Pisces is watery in sympathy, Cancer denoting sweet pure water, Scorpius that which is turbid and Pisces that which is stinking, distasteful and alkaline.

380. SIGNS OF THE SEASONS. Quadrants of the zodiac and signs of the seasons. Aries, Taurus and Gemini are vernal, changeable, govern childhood, the east and the east wind, the first watch of day and night. Cancer, Leo and Virgo are aestival, restful, govern youth, the south and the south wind and the second watch, Libra, Scorpius and Sagittarius are autumnal, changeable, govern adult life, the west and its wind, and the third watch, while Capricorn, Aquarius and Pisces are hibernal, peaceful, govern old age, the north and the north wind and the fourth watch.

The first sign of each season is called tropical as it is the turning point, the second fixed, because when the sun is in it the season is established, and the third bicorporal. Each one of these is related by quartile to the others of its kind, and thus Aries, Cancer, Libra and Capricorn form the tropical tetragone, the indications of which are gentleness, purity and sociability with a tendency to science and details. Then Taurus, Scorpius, Aquarius and Leo form the fixed tetragon, the indications of which are mildness, thoughtfulness and justice, in many cases of litigiousness and pugnacity, and sometimes of endurance in adversity and patience in trouble and injustice. Gemini, Virgo, Sagittarius and Pisces, the bicorporal tetragone, indicate amiability, levity, playfulness, thoughtlessness, discord in business, capriciousness and duplicity.

The influence of the fixed signs according to what has been said is obvious, that of the bicorporal more obscure, and that of the tropical between the two.

We must now turn to the essential characteristics of the planets uncomplicated by any other influence, because the relation of the planets to the signs is such that when they enter them they undergo certain alterations; for the planets like the signs are spiritual forces which change the nature of bodies submitted to their influence, a retrograde planet for example, may change a temperament into a choleric one, or a joyful or anxious one, according as one of the four elements becomes preponderant and alters the activities of the spirit and the conditions.

381. NATURE OF THE PLANETS. The planets always influence whatever is receptive under them. So the results of the action of Saturn are in the direction of extreme cold and dryness, of Jupiter of moderate heat and moisture, of Mars, of extreme heat and dryness, of the sun of not immoderate heat and dryness, less than characterizes Mars, the heat being greater than the dryness. The influence of Venus is towards moderate cold and moisture, the latter predominant, of Mercury towards cold and dryness, the latter rather stronger, which influence however may be altered by association with another star. The moon tends to moderate cold and moisture, the one sometimes dominating the other. For the moon alters in each quarter in accordance with the extrinsic heat it is receiving from the rays of the sun. Comparing it with the seasons of the year, the first week has a spring-like character tending towards warmth and moisture, the second summer-like, warmth and dryness, the third after opposition, autumnal towards cold and dryness, and the fourth winter-like towards cold and moisture. Some people say that moisture always predominates in the moon whatever its station, but as a fact its moisture tends to warmth with the increasing light of the first half and to cold with the decreasing light of the second, because when the extrinsic influence ceases it can only return to its original condition.

382. MALEFICENT AND BENEFICENT. With regard to the good and evil fortune due to the planets, Saturn and Mars are maleficent, the former especially so; Jupiter and Venus are beneficent, especially the former. Jupiter confronts Saturn in clearing-up unfortunate complications as Venus does Mars. The sun is both beneficent and maleficent, the former when in aspect and distant, the latter when in conjunction and near. Mercury also is either very fortunate or the reverse; it assists whatever planet is near it, but when alone is inclined to beneficence, the more so in proportion to its proximity. In virtue of its own nature the moon is fortunate, but its position with regard to the other planets changes quickly owing to the rapidity of its motion.

On the whole the effects of the beneficent planets may be described as virtue, peace, plenty, good disposition, cheerfulness, repose, goodness and learning. If these influences are powerful, they are friendly to each other, if weak, they lend each other assistance. On the other hand, the maleficent effect destruction, tyranny, depravity, covetousness, stupidity, severity, anxiety, ingratitude, shamelessness, meanness, conceit and all kinds of bad qualities. If powerful they help each other in enmity, but if weak, abandon each other, and when alone are active but cowardly.

Some people say that Saturn is at first inimical on account of Mars, and later fortunate on account of Jupiter because it accompanies them in all states. They say of Mars it is at first fortunate and later maleficent, and the same of the sun, but we know of no justification for these ideas, for the principle at the root of this matter is that any planet which has its two qualities in an extreme degree is maleficent; in a moderate degree, beneficent, and that if the qualities are unequally present, then it is neither called beneficent nor maleficent except under certain conditions.

383. EFFECT OF MOON'S NODES. Many astrologers attribute a definite nature to the ascending and descending nodes, saying that the former is warm and beneficent and denotes an increase in all things, and the latter cold, maleficent, and accompanied by a diminution of influences. It is related that the Babylonians held that the ascending node increases the effects of both beneficent and maleficent planets, but it is not every one who will accept these statements, for the analogy seems to be rather farfetched.

384. HINDU OPINION. According to the Hindus, Saturn, Mars, (the sun and the ascending node) are in general maleficent; (they do not mention the Dragon's tail). Jupiter and Venus are in general beneficent, and Mercury increases the effects of both beneficents and maleficents. Of the moon some say that while waxing it is beneficent, and when waning, maleficent, while others assert that for the first ten days it is neither beneficent nor maleficent, during the second ten, beneficent, and during the third, maleficent.

385. MALE AND FEMALE. All the three superior planets and the sun are male, Saturn, among them, being like a eunuch (has no influence on birth). Venus and the moon are female, and Mercury hermaphrodite, being male when associated with the male planets, and female when with the female; when alone it is male in its nature. Some people say that Mars is female, but this opinion is not received.

386. DIURNAL AND NOCTURNAL. Saturn, Jupiter and the sun are diurnal and exercise their power during the day. Mars, Venus and the moon nocturnal and Mercury is either one or the other depending on the sign in which it is, or on the planet with which it is associated. Every planet assists those resembling it, the diurnal asking assistance from the diurnal and the nocturnal from the nocturnal.

The sun is lord of the day and the moon of the night, because their influence is exerted during these periods. Every planet which is under the horizon during its own period is without influence.

Some people say that the dragon's head is male and diurnal and the tail female and nocturnal, but this is quite illogical.

387. ARE INDICATIONS CONSTANT? The indications of a planet do not always remain constant; they are dependent on its relations to the various signs, to other planets and to the fixed stars, to the position as regards the sun and its rays, and to distance from, or proximity to the earth. Thus Saturn which is dry as it rises becomes moist as it sets.

The effects which are thus attributable to the various situations of a planet present themselves in two forms, the one fortunate, the other unfortunate. Saturn, for example, which governs matters of the land, if in conditions of power and beneficence improves the agricultural conditions, blessings and good luck ensue and increased profits are realized; but if the conditions are adverse, the farming operations are attended by disappointment, bad fortune and failure.

All the indications of the planetary influences which are described in the books are set down in the tables which follow.

388. WHY ONE QUALITY REPEATEDLY ATTRIBUTED TO CERTAIN PLANETS AND NOT TO OTHERS. It may be asked why mention is made of several planets in connection with one subject, when the same is not the case with others (the signs). This is due first of all to certain defects in the art, and to confusion of reasoning. The masters of astrology first agreed to arrange things according to their colours, smell, taste, special peculiarities, actions and habits and attached them to planets in accordance with the nature, beneficence or maleficence of these, but other associations were suggested by resemblance in time of appearance or of coming into action. It is rare that only one planet furnishes the indications for one subject or object, generally two or more are associated, as for example when

two elementary qualities are present obviously related to two different planets. Thus the onion is related by its warmth to Mars and by its moisture to Venus, and opium by its coldness to Saturn, and its dryness to Mercury. So when any one speaks of Saturn as the significator of opium, it is merely its coldness that is referred to, and if Mercury is cited in the same capacity, that is due to its dryness. Those people who do not use discrimination in these matters are therefore responsible for the contradictions which occur in their books.

Again there are groups of objects which have as general significator one particular planet, while other planets are associated with the individuals of the group. Thus Venus is the significator for all sweet-smelling flowers, but Mars in the case of the rose is associated with it on account of its thorns, colour and pungent odour which incites catarrh, while Jupiter shares with Venus in the case of the narcissus; Saturn in the case of the myrtle, the Sun in that of the water-lily, Mercury in that of royal basil, and the Moon in that of the violet.

Similarly the various organs of a plant are distributed to different planets. Thus the stem of a tree is appropriated to Sun, the roots to Saturn, the thorns, twigs and bark to Mars, the flowers to Venus, the fruit to Jupiter, the leaves to the moon, and the seed to Mercury. Even in the fruit of a plant like a melon the constituent parts are divided among several planets, the plant itself and the flesh of the fruit belong to the sun, its moisture to the moon, its rind to Saturn, smell and colour to Venus, taste to Jupiter, seed to Mercury and the skin of the seed and its shape to Mars.

389. RELATION TO POINTS OF COMPASS. I have not seen in the ordinary textbooks any reference to a connection between the planets and the points of the compass except in Nayrizi's Book of Nature, who in speaking of the four triplicities refers Saturn to the East, Mars to the West, Venus to the South and Jupiter to the North.

The Hindus, however, attribute to the planets certain powers which they call directional (jihati) this belongs to Mercury and Jupiter at the horoscope, to the sun and moon at the tenth house, to Saturn at the seventh, and to Venus and the moon at the fourth. So it becomes necessary to associate the East with Mercury and Jupiter, the West with Saturn, the South with the sun and Mars and the North with Venus and the moon.

They have also an octagonal figure called ra's which they use in trying to secure victory in gambling. Here they place the sun

Arc of night

Night of 8 equal hours (20 gharis) = 120°

Night of 12 unequal hours 1/12th arc of night = 10°

	12	11	10	9	8	7	6	5	4	3	2	1	12	11	10	9	8	7	6	5	4	3	2
	♃	♀	☉	♂	♃	♄	☽	♀	♀	☉	♂	♃	♄	☽	♀	♀	☉	♂	♃	♄	☽	♀	♀

4 am — Midnight — 8 pm — 4 pm — Noon — 8 am — 4 am

Arc of day

Day of 16 equal hours (40 gharis) = 240° equinoct.

Day of 12 unequal hours 1/12th arc of day = 20°

Night of 12 unequal hours 1/12th arc of day = 20°

Diagram of equal and unequal hours with the Lords of the hours for a Sunday. *See translator's note.*

---

at the East, Jupiter at the South, Mars at the South-East, the moon South-West, Saturn North-West, Mercury North, and Venus North-East, the West point being left vacant.

390. PLANETS AS LORDS OF HOURS AND DAYS OF WEEK. With regard to the distribution of the days of the week among the planets, it is natural that the first hour of the first day Sunday should be given to the planet which is the cause of day and night, viz. the sun. The second hour is allotted to the next lower planet Venus, the third to Mercury, the fourth to the moon, the fifth to Saturn and so on till the second day Monday whose first hour falls to the moon, second to Saturn, and so on in the same way until another Sunday arrives, when the first hour is again the turn of the sun. The lords of the hours having been determined in this way it was natural that the days of the week should be assigned to the planet associated with the first hour thereof.

Some people assert that the odd hours of the twenty-four are male and the even ones female.

391. HINDU DIFFERENCE. The Hindus deal with this matter in a better way. They reckon their day of twenty-four hours from sunrise to sunrise, and

The Arabs divide the whole day from sunrise to sunset into 12 day hours and the night from sunset to sunrise into 12 night hours. In the diagram the day is much longer than the night, 16 of our hours to 8. Therefore Arab hours are very unequal to ours, and are also unequal as the days vary in length. Here 2 spaces are given to Arab day hours and one to night hours, so 3 Arab day hours = 4 of ours and 6 Arab night hours = 4 of ours. [*Translator's note*]

30

allot the whole day to one governing planet without making separate provision for the hours. Their night follows day, and their hours are equal which appears to be the most reasonable method. Our astrologers however, deal separately with day and night and divide them into unequal hours, so that the lord of the night which follows a day is the thirteenth planet counting downwards from the lord of the preceding day and (an easier calculation) the sixth downwards or the third upwards.

It is on account of this that the unequal hours are engraved on the astrolabe, but this method of division is contrary to nature.

392. PLANETS AND CLIMATES. With regard to the seven climates the first from the equator to its boundary is given to Saturn the first and highest planet and the one with the widest orbit, because the first climate is the longest of all, the most generous in yielding the necessities of life, and its inhabitants resemble Saturn in colour and disposition. The second climate belongs to Jupiter and so on to the seventh which is allotted to the moon. Abu Ma'shar regards this as a Persian view, and says that the Greeks give the first climate to Saturn, the second to the sun, the third to Mercury, the fourth to Jupiter, the fifth to Venus, the sixth to Mars and the seventh to the moon.

393. ASCENDANTS OF CITIES, CLIMATES, LORDS OF THEIR HOURS. To find an association between a particular place and a sign or planet is a matter for investigation and research, but how are we to draw a horoscope or ascertain the lord of the hour for a place, unless we know accurately the time of the beginning of its construction? And what city is there of which such a recollection is preserved? Even if a religious ceremony be associated with the foundation of a city, the history of its early conditions has passed into oblivion. Even suppose that is not so, and that we assume a certain date of its foundation, and draw a horoscope and calculate the lords of the hours in accordance therewith, how are we to do so for a well-known stream or a great river, since we can know nothing as to when water began to flow in it and excavate its channel? These questions are futile and their absurdity is obvious to the intelligent.

394. YEARS OF THE PLANETS. The years of the planets are of four degrees, least, mean, great and greatest. The last are only used for marking certain time-cycles, although some people say that in ancient days the planets granted such long years of life. Astrologers of the present day only use the three former degrees for determining the length of life at a nativity, and the numbers which they thus elicit

31

must not be interpreted literally as years, but freely, for sometime they represent years, but sometimes months, weeks, days or hours.

395. FIRDARIA OF PLANETS. The years of a man's life according to a Persian idea are divided into certain periods (firdar) governed by the lords of these known as Chronocrators. When one period is finished another begins. The first period always begins with the sun in a diurnal nativity and with the moon in a nocturnal one; the second with Venus in the one case, in the other with Saturn, the remaining periods with the other planets in descending order. The years of each period are distributed equally between the seven planets, the first seventh belonging exclusively to the chronocrator of the period, the second to it in partnership with the planet next below it and so on.

396-401. NATURES OF THE PLANETS AND THEIR INDICATIONS. The general characteristics of the planets and their indications as to 396 elementary qualities; 397 beneficence or maleficence; 398 sex; 399 whether diurnal or nocturnal; 400 smell and taste; 401 colour.

Saturn is extremely cold and dry. The greater malefic. Male. Diurnal. Disagreeable and astringent, offensively acid, stinking. Jet-black also black mixed with yellow, lead colour, pitch-dark.

Jupiter is moderately warm and moist. The greater benefic. Male. Diurnal. Sweet, bittersweet, delicious. Dust-colour and white mixed with yellow and brown, shining, glittering.

Mars is extremely hot and dry. The lesser malefic. Male (some say female). Nocturnal. Bitter. Dark red.

Sun is hot and dry, the heat predominant. Maleficent when near, beneficent at a distance. Male. Diurnal. Penetrating. Pungent, shining reddish-yellow, its colour is said to be that of the lord of the hour.

Venus is moderately cold and moist, especially the latter. The lesser benefic. Female. Nocturnal. Fat and sweet flavour. Pure white tending to straw-colour, shining, according to some greenish.

Mercury is moderately cold and dry, the latter predominant. Beneficent. Male and diurnal by nature, but takes on the characters of others near. Complex flavour and colour, the latter sky-blue mixed with a darker colour.

Moon is cold and moist, sometimes moderate, changeable. Beneficent and maleficent. Female. Nocturnal. Salt or insipid, somewhat bitter. Blue and white or some deep colour not unmixed with reddish yellow, moderate brilliancy.

402-406. INDICATIONS CONTINUED. Indications as to 402 the properties of things, 403 their form, 404 the days and nights of the week, 405 climates, 406 nature of soils.

Saturn: Coldest, hardest, most stinking and most powerful of things. Shortness, dryness, hardness, heaviness. Saturday (and Wednesday night). First climate. Barren mountains.

Jupiter: Moderate, complete, pleasant, best and easiest things. Moderation, solidity, smoothness. Thursday (and Monday night). Second climate. Easily worked soil.

Mars: Hot, hard, sharp and red things. Length, dryness and coarseness. Tuesday (and Saturday night). Third climate. Waste, hard and stony land.

Sun: Most expert, noble, well-known and generous things. Revolution, mines, worn-outness, empty and vacant places. Sunday (and Thursday night). Fourth climate. Mountains rich in minerals.

Venus: Most pungent, most agreeable and delicious, most beautiful, softest and ripest things. Squareness, dispersion, smoothness. Friday (and Tuesday night). Fifth climate. Soils with abundant water.

Mercury: Mixture of moderate things. Compounded of two things of this nature. Wednesday (and Sunday night). Sixth climate. Sandy soil.

Moon: Thickest, densest, moistest and lightest objects. Density, moisture, opacity, lightness. Monday (and Friday night). Seventh climate. Plains and level ground.

407-408. BUILDINGS AND COUNTRIES. Indications as to, 407, places and buildings; 408, countries.

Saturn: Underground canals and vaults, wells, old buildings, desolate roads, lairs of wild beasts, deserts full of them, stables for horses, asses, and camels, and elephants' houses. India, Zanzibar, Abyssinia, Egypt, Ethiopia between the West and the South,

Yemen, Arabia and Nabatea.

Jupiter: Royal palaces, mansions of the nobility, mosques, pulpits, Christian churches and synagogues, science, books, ordinary vessels, teachers' houses, hamlets of leadworkers. Babylon, Fars, Khurasan, the country of the Teviks and the Berbers in Africa as far as the West.

Mars: (Fire-temples), fireplaces and firewood, roadside fires and the vessels necessary for the art of the potter. Syria, Greece, Slavonia, North-Western countries.

Sun: Kings' and sultan's palaces. Hijjaz, Jerusalem, Mount Lebanon, Armenia, Alan, Dailam, Khurasan as far as China.

Venus: Lofty houses, vessels (roads) which hold much water, places of worship. Babylon, Arabia, Hijaz, and its neighbourhood, (islands and sugar-plantations) and cities of Mesopotamia and the Middle of the Marshes.

Mercury: Bazaars and divans, mosques, houses of painters and bleachers and such as are near orchards, irrigation channels and springs. Mecca, Medina, Iraq, Dilam, Gilan, Tabaristan.

Moon: Moist places, underground or under water brick-making, places to cool water, streams and roads with trees. Mosul, Azarbaijan, the narrow streets of the common people everywhere.

409-411. RELATIONS TO ORES, METALS AND JEWELS, GRAINS AND FRUITS. Indications as to, 409 mines; 410 metals and precious stones; 411, grains and fruit.

Saturn: Litharge, iron slag, hard stones. Lead. Pepper, belleric myrobalan, olives, medlars, bitter pomegranate, lentils, linseed, hempseed.

Jupiter: Marcasite, tutty, sulphur, red arsenic, all white and yellow stones, stones found in ox-gall. Tin, white lead, fine brass, diamond, all jewels worn by man. Wild pomegranate, apple, wheat, barley, rice, durra, chick-peas, sesame.

Mars: Magnetic iron, shadna (lentil-shaped stones) cinnabar, rouge and mosaics. Iron and copper. Bitter almond, seed of turpentine-tree.

Sun: Jacinths, lapis lazuli, yellow sulphur, orpiment, Pharaonic glass, marble, re-algar, pitch. Gold and whatever is coined therefrom for kings. Orange and maize.

Venus: Magnesia and antimony. Silver and gold and jewels set in these, household vessels made of gold, silver and brass, pearls, emeralds, shells. Figs, grapes, dates, origanum and fenugreek.

Mercury: Depilatory, arsenic, amber, all yellow and green stones. All coins struck with name and number such as dinars, dirhams and coppers, old gold and quicksilver, turquoise, coral, tree-coral. Peas, beans, caraway, coriander.

Moon: Nabetean glass, white stones, emerald, moonstone. Silver and things manufactured of silver, such as cups, bangles, rings and the like, pearls, crystal, beads strung. Wheat, barley, large and small cucumbers, melons.

412-413. RELATIONS TO TREES & CROPS

Saturn: Oak-gall tree, citron or myrobalan tree, olive tree and also willow, turpentine tree, castor-oil, plant, and all those which bear fruits with disagreeable taste or smell, or hard-shells such as walnuts and almonds. Sesame.

Jupiter: Trees bearing sweet fruit without hard skin such as peach, fig, apricot, pear and lote-fruit, companions Venus as to fruits. Roses, flowers, herbs sweet-smelling or tall, such plants as are light and whose seeds fly with the wind.

Mars: All bitter, pungent and thorny trees, their fruit with rough skin, pungent or very bitter such as bitter pomegranate, wild pear, bramble. Mustard, leeks, onions, garlic, rue, rocket, wild rue, radish, eggplant.

Sun: All tall trees which have oily fruit, and those whose fruit is used dry, such as date-palms, mulberries and vines. Dodder, sugarcane, manna, tarangubin and shir-khisht.

Venus: All trees soft to touch, sweet-smelling, smooth to the eye like cypress and teak, apple and quince. Sweet and oily berries, fragrant and coloured herbs, spring flowers and has a share in cotton.

Mercury: Pungent and evil-smelling trees. Savoury herbs and garden stuff, canes and things growing in water.

Moon: All trees the stem of which is short such as the vine and the sweet pomegranate. Grass, reeds, canes, flax, hemp, trailing plants such as cucumber and melon.

414-417. RELATIONS TO FOODS & DRUGS, HOUSEHOLD REQUIREMENTS, STATES OF BEING, POWERS. Indications as to, 414, foods and drugs; 415, household utensils; 416, states of being, 417, powers.

Saturn: Drugs cold and dry in the fourth degree, especially those which are narcotic and poisonous. Dwellings. Sleep. Retentive power.

Jupiter: Those which are moderately hot and moist and are profitable and agreeable. Fruits. Clothing. Vital, growing nutritive faculties and the air in the heart.

Mars: Whatever is not poisonous but pungent and warm in the fourth degree. Drugs. Business. Passion.

Sun: Whatever is warm beyond the fourth degree and is salutary and in general use. Foods. Eating and drinking. Youthful vigour.

Venus: Moderately cold and moist foods, useful and pleasant to the taste. Savoury herbs. Coition. Sensuality.

Mercury: Foods which are dryer than cold and are agreeable but rarely useful. Grains. Speaking. Faculty of reflection.

Moon: Foods which are equally cold and moist, sometimes useful, sometimes detrimental, and are not in constant use. Beverages. Drinking water. Natural power.

418. INDICATIONS AS TO QUADRUPEDS &c.

Saturn: Black animals and those living in holes in the ground; oxen, goats, horses, sheep; ermine, sable, weasel, cat, mouse, jerboa, also, large black snakes, scorpions and other poisonous insects and fleas and beetles.

Jupiter: Man, domestic animals and those with cloven hoofs such as sheep, oxen, deer, those which are speckled and beautifully coloured, and edible, or speaking, or trained such as lions, cheetahs and leopards.

Mars: Lion, leopard, wolf, wild pig, dog, destructive or mad wild beasts, venomous serpents.

Sun: Sheep, mountain goat, deer, Arab horse, lion, crocodile, nocturnal animals which remain concealed during the day.

Venus: All those wild animals which have white or yellow hoofs such as gazelle, wild ass, mountain goat also large fish.

Mercury: Ass, camel, domestic dog, fox, hare, jackal, ermine, nocturnal creatures, small aquatic and terrestrial animals.

Moon: Camel, ox, sheep, elephant, giraffe, all beasts of burden obedient to man and domesticated.

419-422. INDICATIONS AS TO BIRDS, ELEMENTS AND HUMOURS, ORGANS OF THE BODY, VITAL ORGANS. Indications as to, 419, birds and other fliers; 420, the elements and the humours; 421, organs of similar nature; 422, vital organs.

Saturn: Aquatic and nocturnal birds, ravens, swallows and flies. Earth, black bile and occasionally crude phlegm. Hair, nails, skin, feathers, wool, bones, marrow and horn. Spleen.

Jupiter: Birds with straight beaks, grain eating, not black, pigeon, francolin, peacock, domestic fowls, hoopoo and lark. Air and blood. Arteries, sperm and bone marrow. Heart in partnership with the sun.

Mars: Flesh-eating birds with curved bills, nocturnal, water hens, bats, all red birds, wasps. The upper part of fire and yellow bile. Veins and the hinder regions. Liver together with Venus.

Sun: Eagle, ring-dove, turtle dove, cock and falcon. The lower part of fire. Brains, nerves and the hypochondria, fat and everything of this kind. Stomach.

Venus: Ring-dove, wild pigeon, sparrow, bulbul, nightingale, locusts and inedible birds. ——. Flesh, fat and spinal marrow. Kidneys.

Mercury: Pigeon, starling, crickets, falcon, aquatic birds and nightingales. Black bile. Arteries. Gall-bladder.

Moon: Ducks, cranes, carrion, crows, herons, chicks, partridge. Phlegm. Skin and everything related thereto. Lungs.

423-426. INDICATIONS AS TO PARTS OF HEAD, SENSES, MEM-
BERS OF BODY, TIME OF LIFE. Indications as to, 423, parts of the
head; 424, sense organs; 425, paired and other organs; 426, period
of life

Saturn: Right ear. Hearing. Buttocks, podex, bowels, pe-
nis, back, height, knees. Old age.

Jupiter: Left ear. Hearing and touch. Thighs and intes-
tines, womb and throat. Middle age.

Mars: Right nostril. Smell and touch. Legs, pubes, gall-
bladder, kidney. Youth.

Sun: Right eye. Sight. Head and chest, sides, teeth, mouth.
Full manhood.

Venus: Left nostril. Smell and inhaling organs. Womb,
genitals, hands and fingers. Youth and adolescence.

Mercury: Tongue together with Venus. Taste. Organs of
speech. Childhood.

Moon: Left eye. Vision and taste. Neck, breasts, lungs, stomach, spleen. Infancy to old age according to its various quarters.

427-428. RELATIONS AND CONNECTIONS, FIGURE AND FACE

Saturn: Fathers, grandfathers, older brothers and slaves.
Ugly, tall, wizened, sour face, large head, eyebrows joined, small eyes,
wide mouth, thick lips, downcast look, much black hair, short neck,
coarse hand, short fingers, awkward figure, legs crooked, big feet.

Jupiter: Children and grandchildren. Fine figure, round
face, thick prominent nose, large eyes, frank look, small beard, abundant curly hair reddish.

Mars: Brothers of middle age. Tall, large head, small eyes
and ears, and fine forehead, sharp grey eyes, good nose, thin lips,
lank hair, reddish, long fingers, long steps.

Sun: Fathers and brothers, slaves. Large head, complexion
white inclining to yellow, long hair, yellow in the white of the eye,
stammers, large paunch with folds.

Venus: Wives, mothers, sisters, uterine kindred, delicate child. Fine round face, reddish-white complexion, double chin, fat cheeks, not too fat, fine eyes, the black larger than the white; small teeth, handsome neck, medium tall, short fingers, thick calves.

Mercury: Younger brothers. Fine figure, complexion brown with a greenish tinge, handsome, narrow forehead, thick ears, good nose, eyebrows joined, wide mouth, small teeth, thin beard, fine long hair, well-shaped long feet.

Moon: Mothers, maternal aunts, elder sisters, nurses. Clear white complexion, gait and figure erect, round face, long beard, eyebrows joined, teeth separate crooked at the points, good hair with locks.

429. DISPOSITION AND MANNERS.

Saturn: Fearful, timid, anxious, suspicious, miserly, a malevolent plotter, sullen and proud, melancholy, truth-telling, grave, trusty, unwilling to believe good of anyone, engrossed in his own affairs and consequently indicates discord, and either ignorance or intelligence, but the ignorance is concealed.

Jupiter: Good disposition, inspiring, intelligent, patient, high-minded, devout, chaste, administering justice, truth-telling, learned, generous, noble, cautious in friendship, egoistic, friend of good government, eager for education, an honourable trusty and responsible custodian, religious.

Mars: Confused opinions, ignorant, rash, evil conduct, licentious, bold, quarrelsome, unsteady, untrustworthy, violent, shameless, unchaste but quickly repentant, a deceiver, cheerful, bright, friendly and pleasant-faced.

Sun: Intelligent and knowledgeful, patient, chaste, but sensual, eager for knowledge, power and victory, seeking a good name for helping others, friendly, hot-tempered but quickly recovering repose.

Venus: Good disposition, handsome face, good-natured inclined to love and sensuality, friendliness, generosity, tenderness to children and friends, pride, joy, patience.

Mercury: Sharp intelligence and understanding, affability, gentleness, open countenance, elegance, far-sightedness, changeable, deeply interested in business, eager for pleasure, keeps secrets, seeking friendship of people, longing for power, reputation and approval, preserves true friends and withdraws from bad ones, keeps away from trickery, strife, malevolence, bad-heartedness and discord.

Moon: Simple, adaptable, a king among kings, a servant among servants, good-hearted, forgetful, loquacious, timid, reveals secrets, a lover of elegance, respected by people, cheerful, a lover of women, too anxious, not intellectually strong much thought and talk.

430. ACTIVITIES, INSTINCTS AND MORALS.

Saturn: Exile and poverty, or wealth acquired by his own trickery or that of others, failure in business, vehemence, confusion, seeking solitariness, enslaving people by violence or treachery, fraud, weeping and wailing and lamentation.

Jupiter: Friendliness, a peacemaker, charitable devoted to religion and good works, responsible, uxorious, laughing, eloquent, eager for wealth, in addition to affability some levity and recklessness.

Mars: Marriage, travelling, litigation, business going to ruin, false testimony, lustful, a bad companion, solitary, spiteful and tricky.

Sun: Longing for power and government, hankering after wealth and management of worldly affairs, and imposing will on the ignorant, reproving evil-doers, harsh with opponents. If sun is in exaltation, the position is favourable to kings, if in fall to those in rebellion.

Venus: Lazy, laughing, jesting, dancing, fond of wine, chess, draughts, cheating, takes pleasure in every thing, not quarrelsome, a sodomite or given to excessive venery, well-spoken, fond of ornaments, perfume, song, gold, silver, fine clothes.

Mercury: Teaching manners, theology, revelation and logic, eloquent, fine voice, good memory for stories, ruining prospects by too great anxiety and misfortunes, fearful of enemies, frivolous, eager to buy slaves and girls, busybody, calumnious, thieving, lying and falsifying.

Moon: Lying, calumniation, overanxious for health and comfort, generous, in distributing food, too uxorious, levity in appropriate places, excellent spirits.

Saturn: Sickness, affliction, poverty, death, disease of internal organs, gout. Owners of estates, kings' intendants, religious of various sects, devotees, wicked people, bores, the overworked, eunuchs, thieves, the moribund, magicians, demons, ghouls, and those who revile them.

Jupiter: Sickness, fatigue, fever, death in childbed, Caesarean section. Kings, vazirs, nobles, magnates, lawyers, merchants, the rich and their sycophants.

Mars: Fever. Leaders, cavalry, troops, opponents, disputants in assembly.

Sun: ——. Kings, nobles, chiefs, generals, officials, magistrates, physicians, societies.

Venus: ——. Nobles, plutocrats, queens, courtesans, adulterers and their children.

Mercury: ——. Merchants, bankers, councillors, tax-collectors, slaves and wrestlers.

Moon: Diseases of many kinds. Kings, nobles, noble matrons; celebrated, and wealthy citizens.

433-434. INDICATIONS AS TO RELIGIONS; PICTURES OF PLANETS. Indications as to, 433, religions; 434, pictorial representations of the planets.

Saturn: Jews and those who dress in black. Old man seated on a wolf, in his right hand the head of a man and in the left a man's hand; or according to another picture, mounted on a bright bay horse, on his head a helmet, in the left hand a shield and in the right a sword.

Jupiter: Christians and those dressed in white. A young man with a drawn sword in the right hand and a bow and a rosary in the left, on horseback; another picture: man on a throne, clad in variously coloured robes, a rosary in the left hand.

Mars: Idolaters, wine-bibbers, dressed in red. Young man seated on two lions, in the right hand a drawn sword in the left a battle-axe; another picture: mounted on a bay horse, helmet on head, in the left hand a spear adorned with red roses, pennon flag, in the right hand head of a man, clad in red.

Sun: Wearing a crown; Magians, Mithraists. A man seated on something like a shield on wheels drawn by four oxen, in his right a staff on which he rests, in his left a mace gurz, beads; another picture: man (jurs) seated, face like a circle, holding reins of four horses.

Venus: Islam. Woman on a camel holding a lute which she is playing; another picture: woman seated her hair unloosened the locks in her left hand, in the right a mirror in which she keeps looking, dressed in yellowish green, with a necklace, bells, bracelets and anklets.

Mercury: Disputants in all sects. Youth seated on a peacock, in his right hand a serpent and in the left a tablet which he keeps reading; another picture: man seated on a throne, in his hand a book which he is reading, crowned, yellow and green robe.

Moon: Adherents of the prevailing religion. Man with javelin in right hand, in his left thirty, you would think there were three hundred, on his head a crown, seated in a chariot drawn by four horses.

435. INDICATIONS AS TO TRADES, PROFESSIONS, ETC.

Saturn: Building, paymaster, farming, reclaiming land and distribution of water, (fraudulent transactions), apportioning money and heritages, grave-digging; selling things made of iron, lead, bone, hair; copper, black slaves; knowledge used for bad purposes, such acts of the government as lead to evil oppression, wrath, captivity, torture.

Jupiter: Noble actions, good government, religion, doing good; interpretation of dreams; goldsmiths' work, banking; selling old gold and silver, white clothes, grapes and sugarcane.

Mars: Lawmaking, selling and making armour, blacksmiths craft, grooms, shepherds, butchers, veterinary surgeons, surgeons, circumcisers, sellers of hounds, cheetahs, boars, wolves, copper, sickles, beer, glass, boxes, wooden cups, brigandage, contention, housebreaking, highwaymen, grave-robbers and prison, torture, execution.

Sun: Receiving, giving and selling gold-brocades.

Venus: Works of beauty and magnificence, fond of bazaars, commerce, measuring by weight, length and bulk; dealing in pictures and colours, goldsmiths work, tailoring, manufacturing perfumes, dealing in pearls, gold and silver ornaments, musk, white and green clothes, maker of crowns and diadems, accompanying singing, composing songs, playing the lute, feasts, games and gaming.

Mercury: Merchants, calculators and surveyors, astrologers, necromancers and fortune-tellers, geometrician, philosopher, disputation, poetry, eloquence, manual dexterity and anxiety for perfection in everything, selling slaves, hides, books, coins; profession of barber, manufacture of combs.

Moon: Engaged in business matters, missions, agencies, accounting; strenuous in religion and divine law, skill in all branches; practice of medicine, geometry, the higher sciences, measuring land and water; growing and cutting hair; selling food, silver rings and virgins, also indicates captivity, and prison for the deceptions of wizards.

436-437. ORBS AND YEARS OF PLANETS

Orbs		Years			
		least	mean	great*	greatest
Saturn	9^0	30	43½	57	265
Jupiter	9^0	12	45½	79	427
Mars	8^0	15	40½	66	284
Sun	15^0	19	39½	120	1461 (sothiac cycle)
Venus	7^0	8	45	82	1151
Mercury	7^0	20	48	76	461
Moon	12^0	25	39½	108	520

* The great years are the sums of the Egyptian Ptolemaic terms of each planet; the least of Saturn and moon have been related to their periods of revolution, of Sun to the Metonic cycle, of Venus to its orb, while those of Mars and Mercury and the greatest years remain unexplained. In the case of Sun and Moon, the mean is: $\frac{least + ½ great}{2}$ but cf. Vet. Val. p. 157 and B.L. 410 where Sun and Moon treated like other planets. Vettius Valens p. 164 has another explanation for the great years of the planets:

ħ ¼ of ☉ great years + ¼ great ☽ = 57
♃ ½ of ☉ great years + least = 79
  or ½ of ☽ great years + least = 79
♂ ½ of ☽ great years + least of ♃ = 66
♀ ½ of ☽ great years + least of ħ = 84 (the sum of the terms = 82)
☿ ½ of ħ great years + least of ☉ = 76

43

438-439. FIRDARIA AND THEIR ASSOCIATION TIMES. Periods of life controlled by the planets as chronocrators, 438, and the times of association, 439, (sevenths of the periods) of the other planets with the general chronocrators.

	Diurnal Nativities		Nocturnal Nativities	
Periods*		Time of association in last six sevenths		Time of association in last six sevenths
1	☉ 10 years	1 yr. 5 m. 4 d. 7 h.	☽ 9 years	1 yr. 3 m. 12 d. 21 h.
2	♀ 8 years	1 yr. 1 m. 21 d. 5 h.	♄ 11 years	1 yr. 6 m. 25 d. 17 h.
3	☿ 13 years	1 yr. 10 m. 8 d. 7 h.	♃ 12 years	1 yr. 8 m. 17 d. 7 h.
4	☽ 9 years	1 yr. 3 m. 12 d. 21 h.	♂ 7 years	1 yr. 10 h.
5	♄ 11 years	1 yr. 6 m. 25 d. 17 h.	☉ 10 years	1 yr. 5 m. 4 d. 7 h.
6	♃ 12 years	1 yr. 8 m. 17 d. 3 h.	♀ 8 years	1 yr. 1 m. 21 d. 5 h.
7	♂ 7 years	1 yr.	☿ 13 years	1 yr. 10 m. 8 d. 7 h.
8	☊ 3 years day & night *(no association)*		☋ 2 years day & night *(no association)*	

440. DOMICILES OF THE PLANETS. We now proceed to discuss the relation of the planets to the signs.

The zodiac belt is divided into two halves, the first extending from the beginning of Leo to the end of Capricorn, and this half is given to the sun whose domicile is the first sign, viz. Leo. The other half is given to the moon; it extends from the beginning of Aquarius to the end of Cancer in which sign its domicile is. As the other planets have two methods of movement retrograde and direct, so also they have each two domiciles one on the sun side and one on the moon side, at equal distances from the interval between Leo and Cancer. Beginning with Mercury the nearest planet, Virgo on the sun side and Gemini on the moon side are assigned to it as domiciles, then Libra and Taurus to Venus, Scorpius and Aries to Mars, Sagittarius and Pisces to Jupiter and Capricorn and Aquarius to Saturn as in the annexed figure (*next page*).

441. DOMICILES PREFERRED. One of these domiciles is always more congenial to the planets and it is said that there they are more Joyful on account of temperament, formation, and sex. The sun and moon, however, as they are not confined to one domicile find conditions in all. But of those which have two, Mercury prefers Virgo to Gemini, Venus Taurus, Mars Aries, Jupiter Sagittarius, Saturn Aquarius.

---

* A span of 75 years, by day or night, is thus provided for.

440. Domiciles of the planets.

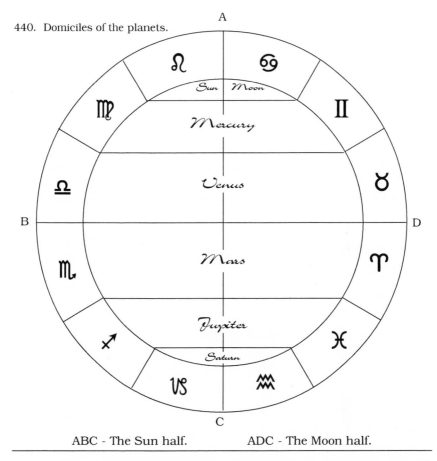

ABC - The Sun half.   ADC - The Moon half.

The opinion of the Hindus agrees in some respects and differs in others; they say that Mars finds Aries more congenial, the moon Taurus, the sun Leo, Mercury Virgo, Venus Libra, Jupiter Sagittarius, and Saturn Aquarius. They call such situations mulatrikuna and assert that a planet occupying one of these has more influence than in its own domicile.

442. DETRIMENTS. The signs opposite to the domiciles of the planets are said to be their detriments or debilities. The Hindus while recognizing the domiciles do not know this expression. The detriments are shown in the accompanying figure (*next page*).

443. EXALTATION AND FALL OF PLANETS. There are certain signs which are described as places of exaltation of the planets, like the thrones of kings and other high positions. In such signs the exaltation is regarded as specially related to a certain degree, but there are many differences of opinion in this matter, some saying that it

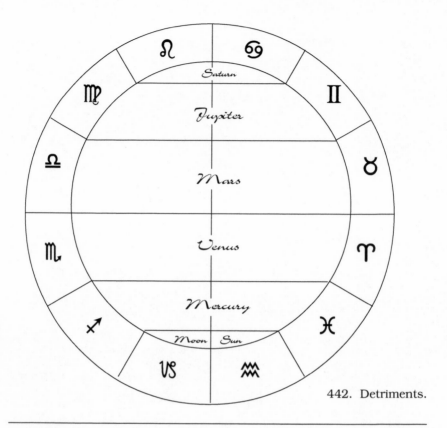

442. Detriments.

extends to some degrees in front of or behind the degree in question, while others hold that it extends from the first point of the sign to that degree, and again others that it is present in the whole sign without any special degree. Below are the signs and degrees according to the Persians and Greeks.

Saturn	21⁰	of Libra
Jupiter	15⁰	of Cancer
Mars	28⁰	of Capricorn
Sun	19⁰	of Aries
Venus	27⁰	of Pisces
Mercury	15⁰	of Virgo
Moon	3⁰	of Taurus
Dragon's Head	3°	of Gemini
Dragon's Tail	3°	of Sagittarius

The opposite signs and degrees are regarded as places of dejection for the planets, when in them, they are said to be in their fall, and are therein confined and their condition deteriorated.

46

444. HINDUS DIFFER AS TO DEGREES. There is no difference of opinion as to the signs of exaltation, but the Hindus differ as to the degrees in certain cases. They are agreed that the exaltation of the sun lies in 10^0 of Aries, of Jupiter in 5^0 of Cancer, of Saturn 20^0 of Libra, the others as above, except with regard to the Dragon's Head and Tail which are not mentioned by them in this connection as is quite proper.

445. LORDS OF HIS TRIPLICITIES. Each triplicity has a lord by day and another by night, also a third which shares this responsibility both by day and night. Thus the fiery triplicity has as lord the sun by day, and Jupiter by night, while Saturn is a partner both by day and night. The earthly triplicity has Venus by day, the moon by night, Mars being in this case the partner. The airy triplicity has Saturn by day, Mercury by night and Jupiter as partner, while the watery triplicity has Venus by day, Mars by night and the moon as partner.

Their Lords

The Triplicities				by Day		by Night	
1st Fiery	♈	♌	♐	☉	♄	♃	♄
2nd Earthy	♉	♍	♑	♀	♂	☽	♂
3rd Airy	♊	♎	♒	♄	♃	☿	♃
4th Watery	♋	♏	♓	♀	☽	♂	☽

However Hashwiyite* astrologers associate all three planets at the same time with each triplicity, and merely make the following distinction between day and night, e.g. the lords of the fiery triplicity are the Sun, Jupiter and Saturn by day, and Jupiter, the Sun and Saturn by night and the rest on this analogy. They do not desert their position on consideration, but have filled their books with decrees based thereon, and propositions deduced from these.

446. ASPECTS OF PLANETS IN SIGNS. Whenever two planets are in signs which are in aspect to each other, they also are said to be in aspect; if they are in the same sign they are described as conjunct, while if they are at the same degree the conjunction is said to be partile. If one of them is in a sign third from the other, they are in sextile aspect to the right or left, if in a fourth sign, to be in quartile, if in a fifth in trine, and if in the seventh, opposite. Should

*Al-Biruni had a poor opinion of the Hashwiyites - v. Chron. p. 90, and 527 and 529. As to their doctrine, a creed of the common people as compared with the more aristocratic and intellectual Mutazilites, and the origin of the name v. Van Vloten - Hashwiya et Nabita. Inter. M. Congr. 1897 and Goldziher - Livre de Ibn Toumert. p. 65. Alger 1905. Dict. sci. terms p. 396.

their degrees be equal they are styled muttasilin for then between these aspects it is possible to construct either a regular hexagon, or a square or a triangle in the zodiac, or to divide it into two.

447. FRIENDSHIP AND ENMITY OF PLANETS. Friendship or enmity between the planets is, according to us, based on what we have said as to their domiciles, but astrologers have different theories on this matter. There are those who base them on the temperament and nature of the planets themselves, Saturn and Jupiter being regarded as inimical because the one is dark, maleficent and extremely distant, while the other is shining, beneficent and only moderately distant. There are others who base them on their elementary qualities, those that are fiery being inimical to the watery, and the airy to the earthy, while there are still others who found them on the relative situations of their domiciles and exaltations, if the aspect of these is inimical then their lords are also inimical: further any planet whose domicile is twelfth from the house occupied by another planet is inimical to the latter. When the basis of enmity is arrived at in any of the ways we have enumerated, then that for friendship and indifference becomes obvious.

The views of Abu 'l-Qasim, the philosopher, based on the foregoing considerations are shown in the columns of the subjoined table.

Planet	mutually hurtful with	injurious to	offering friendship to	asking friendship from
Saturn	Sun & Moon	Jupiter	Mars	Venus
Jupiter	Mars Mercury	Mercury	Venus	Moon
Mars	Jupiter Venus	Moon	Sun	Saturn
Sun	Saturn	Venus	-	Mars
Venus	Mars Mercury	-	Saturn	Jupiter
Mercury	Jupiter Venus	Venus	neither offers nor asks friendship	
Moon	Saturn	Mars	Jupiter	Venus

The astrologers of our day however, lay little stress on the friendship or enmity of the planets in the matter of judicial astrology. The Hindus on the other hand regard them as equally important or more so than the domiciles and exaltations, we have accordingly set down their opinions in the accompanying table.

Planets	Friends	Enemies	Indifferent
☉	♃ ♂ ☽	♄ ♀	☿
☽	☉ ☿	*(none)*	♄ ♃ ♂ ♀
♂	♃ ☉ ☽	☿	♄ ♀
☿	☉ ♀	☽	♄ ♃ ♂
♃	♂ ☉ ☽	♀ ☿	♄
♀	♄ ☿	☉ ☽	♃ ♂
♄	♀ ☿	♂ ☉ ☽	♃

As far as friendship or enmity is concerned, they are liable to change, because if a planet meets another in the 10th, 11th, 12th, 2nd, 3rd, or 4th houses,* if friendly the friendship becomes complete, if indifferent becomes friendly, and if inimical indifferent. Also if it meets another in any of the other houses, the effects are precisely the reverse of these.

448. HALF-SIGN. We shall now speak of the different parts of the signs and the fate of the planets therein.

Half of a sign is called 'hour' by the Hindus. The first half of every male sign belongs to the sun and the second to the moon, and on the contrary, of every female sign the first half belongs to the moon and the second to the sun. My friends, in this matter continue to obtain conclusions which differ from the above or are directly opposed thereto, and indeed the distinction between the two cannot be compared with that between light and darkness, as we have said and shall continue to say, but the people who have made use of this distinction are agreed upon its value, in spite of the opinions of others.

449. FACES. Each third of a sign - ten degrees - is called a face and the lords of these faces according to the agreement of the Persians and Greeks are as follows: The lord of the first face of Aries is Mars, of the second the sun, of the third Venus; of the first of Taurus, Mercury and so on in the order of the planets from above downwards till the last face of Pisces.

450. FIGURES. The so-called 'figures' are in reality also the faces, but called so because the Greeks, Hindus and Babylonians associated with each face as it arose the figure of a personage human or divine, and in the case of the Greeks the faces were also associated with such of the other 48 constellations ascending at the same time. But this duplication of constellations is mentioned in connection

* Cf. the statement India II 224, where a planet's nature is said to undergo a change towards friendliness in the Eastern and towards enmity in the Western houses without reference to meeting another planet there.

with affairs, designs and undertakings which are peculiar to the country in question, and is used to obtain decrees with regard to these. We shall not undertake to give an account of it both to save space, and because it would be useless, as the astrological books we have are destitute of any instructions for using it.

451. DECANATES. By the Hindus these thirds of a sign are called darigan or Drikan (decanate), but their lords are different from those of the faces, because the first decanate has as lord the lord of the whole sign, the second, the lord of the fifth sign from it, and the third, the lord of the ninth sign. The lords of the faces and of the Hindu decanates are set down in the table.

Signs	Lords of Faces			Hindu Lords of Decanates		
	10°	20°	30°	10°	20°	30°
♈	♂	☉	♀	♂	☉	♃
♉	☿	☽	♄	♀	☿	♄
♊	♃	♂	☉	☿	♀	♄
♋	♀	☿	☽	☽	♂	♃
♌	♄	♃	♂	☉	♃	♂
♍	☉	♀	☿	☿	♄	♀
♎	☽	♄	♃	♀	♄	☿
♏	♂	☉	♀	♂	♃	☽
♐	☿	☽	♄	♃	♂	☉
♑	♃	♂	☉	♄	♀	☿
♒	♀	☿	☽	♄	☿	♀
♓	♄	♃	♂	♃	☽	♂

452. PTOLEMY'S SIGN-THIRDS. Ptolemy has also made use of the thirds of the signs. He determined by experience and observation of the signs the changes in the atmosphere which are indicated by the signs as a whole, by the individual thirds in these in longitude and by their northern and southern parts in latitude. So whenever the action of the planets on the weather and of their situations at the times of conjunction and opposition in longitude and latitude when weather prognostics are sought it is not easy to estimate the combined effect of all of these influences, as well as of the association and separation of the planets and the fixed stars. The following table is taken from Ptolemy.

	Indications of					
	Whole sign	*North part*	*South part*	*1st third*	*2nd third*	*3rd third*
♈	Thunder & rain	Bringing heat & destruction	Bringing cold & ice	Wind, rain & thunder	Temperate	Burning hot plague epidemics
♉	Heat inclining to moisture	Temperate	Unsettled condition	Earthquakes & hot winds	Cold & wet	Heat, lightning, thunderbolts
♊	Temperate	Winds drying up ground	Scorching heat	Destructive moisture	Temperate	Unsettled
♋	Improvement warm	Scorching heat	Scorching heat	Hot winds & earthquakes	Temperate	Winds
♌	Heat	Wind	Moisture	Hot depressing atmosphere	Temperate	Destructive moisture
♍	Moisture & thunder	Wind	Temperate	Very hot & destructive	Temperate	Very wet
♎	Changeable	Great heat	Moisture bringing epidemics	Fine weather	Temperate	Very wet
♏	Thunder & lightning	Wind	Moisture	Snow & wind	Temperate	Earthquakes
♐	Windy	Wind	Very wet & unsettled	Moisture	Temperate	Very hot
♑	Very wet	Very wet bringing destruction	Very wet & changeable	Great heat & destruction	Temperate	Rains
♒	Cold and wet	Great heat	Wind & snow	Very wet	Temperate	Winds
♓	Cold and wet	Wind	Wet	Moderate	Very wet	Very hot

453. TERMS. These are unequal divisions of the signs known as terms, with each one of them a planet is associated. People however differ in this matter, some holding to the Chaldean, i.e. the ancient Babylonian method, others to that of Astaratus, while others again adopt the scheme of the Hindus. None of these are employed by professional astrologers, who are unanimous in using the Egyptian terms, because they are more correct. Those who have expounded Ptolemy's works use the terms which he records having found in an old book, and which he has inserted in his **Tetrabiblos**. We have constructed a table showing both the Egyptian and the Ptolemaic terms: there is no use discussing any others.

Signs	Egyptian Lords of Terms[1]					Ptolemy's Lords of Terms[2]				
♈	♃ 6	♀ 12	☿ 20	♂ 25	♄ 30	♃ 6	♀ 14	☿ 21	♂ 26	♄ 30
♉	♀ 8	☿ 14	♃ 22	♄ 27	♂ 30	♀ 8	☿ 15	♃ 22	♄ 25	♂ 30
♊	☿ 6	♃ 12	♀ 17	♂ 24	♄ 30	☿ 7	♃ 13	♀ 20	♂ 26	♄ 30
♋	♂ 7	♀ 13	☿ 19	♃ 26	♄ 30	♂ 6	♃ 13	☿ 20	♀ 27	♄ 30
♌	♃ 6	♀ 11	♄ 18	☿ 24	♂ 30	♄ 6	☿ 13	♀ 19	♃ 25	♂ 30
♍	☿ 7	♀ 17	♃ 21	♂ 28	♄ 30	☿ 7	♀ 13	♃ 18	♄ 24	♂ 30
♎	♄ 6	☿ 14	♃ 21	♀ 28	♂ 30	♄ 6	♀ 11	♃ 19	☿ 24	♂ 30
♏	♂ 7	♀ 11	☿ 19	♃ 24	♄ 30	♂ 6	♃ 12	♀ 21	☿ 24	♂ 30
♐	♃ 12	♀ 17	☿ 21	♄ 26	♂ 30	♃ 8	♀ 14	☿ 19	♄ 25	♂ 30
♑	☿ 7	♃ 14	♀ 22	♄ 26	♂ 30	♀ 6	☿ 12	♃ 19	♂ 25	♄ 30
♒	☿ 7	♀ 13	♃ 20	♂ 25	♄ 30	♄ 6	☿ 12	♀ 20	♃ 25	♂ 30
♓	♀ 12	♃ 16	☿ 19	♂ 28	♄ 30	♀ 8	♃ 14	☿ 20	♂ 26	♄ 30

454. HINDU TERMS. The Hindus use only one series of terms for all the male signs, and the same series in the inverse direction for the female signs. This is called their trishanash, or the divisions of the thirty degrees. The result of the arrangement is that the division of the sign is not the same in the two sets, and consequently when it is desired to know which term applies, it is necessary to reckon it out. The series is shown in the annexed table as reported to us -

---

[1] Differs from that given by Firmicus Maternus.
[2] Differs from that given by William Lilly - *Publisher's notes.*

Terms of male signs from the beginning a/c to the Hindus →	5	5	8	7	5	Terms of female signs from the beginning a/c to the Hindus ←
	♂	♄	♃	☿	♀	

455. NINTHS OF THE SIGNS. The Hindus regard the ninth part of a sign, 3°20', which they call nuvanshaka, as very important. When a planet is in its own domicile and ninth, that ninth is called 'bargutam' or most important. The table shows the ninths of all the signs; the lords of the ninths are the lords of the signs concerned. The first ninth of the tropical signs, the fifth of the fixed and the ninth of the bicorporal ones are called 'bargutam' (vargottama). This is an entirely Hindu method on which all are agreed. My friends have altered the order of the lords of the ninths and have arranged them in the order of the spheres, but it is better that we abstain from using it.

		♐ < ♌ / ♈	♑ < ♍ / ♉	♒ < ♎ / II	♓ < ♏ / ♋	
1st	3°20'	Aries / Mars	Capricorn / Saturn	Libra / Venus	Cancer / Moon	Tropical signs
	6°40'	Taurus / Venus	Aquarius / Saturn	Scorpio / Mars	Leo / Sun	Fixed
	10°	Gemini / Mercury	Pisces / Jupiter	Sagittarius / Jupiter	Virgo / Mercury	Bicor- poral
	13-20'	Cancer / Moon	Aries / Mars	Capricorn / Saturn	Libra / Venus	Tropical
5th	16°40'	Leo / Sun	Taurus / Venus	Aquarius / Saturn	Scorpio / Mars	Fixed
	20°	Virgo / Mercury	Gemini / Mercury	Pisces / Jupiter	Sagittarius / Jupiter	Bicor- poral
	23°20'	Libra / Venus	Cancer / Moon	Aries / Mars	Capricorn / Saturn	Tropical
	26°40'	Scorpio / Mars	Leo / Sun	Taurus / Venus	Aquarius / Saturn	Fixed
9th	30°	Sagittar. / Jupiter	Virgo / Mercury	Gemini / Mercury	Pisces / Jupiter	Bicor- poral

The 1st 5th and 9th of these columns form respectively the fiery, earthy, airy and watery triplicities.

456. TWELFTHS OF THE SIGNS. A sign may also be divided into twelfths of 2°30', each of which has a lord, the first twelfth having as lord the lord of the whole sign, the second, the lord of the next sign in succession, and so on to the end of the series. As multiplication is easier than division, and it is difficult for any one to subtract by 2½ degrees, people simplify the calculation by multiplying the number of degrees and minutes of the particular twelfth, the lord of which one wishes to know, by twelve, and then for every 30° counting one sign in the direction of succession from that in which the twelfth is; the last complete 30° indicates the sign whose lord is the lord of the twelfth in question. The lords of the various twelfths of the signs are shown in the table.

This is the division as to which the Greeks and the Hindus are in agreement, but I have always been surprised that my friends have not altered it according to the succession of the signs or some other scheme. For if you proceed according to such a Method you do not commit other absurdities to mention which this is not the place.

Twelfths of the Signs

Lords of the 1/12ths	♈	♉	♊	♋	♌	♍	♎	♏	♐	♑	♒	♓
♂	1st	12th	11th	10th	9th	8th	7th	6th	5th	4th	3rd	2nd
♀	2nd	1st	12th	11th	10th	9th	8th	7th	6th	5th	4th	3rd
☿	3rd	2nd	1st	12th	11th	10th	9th	8th	7th	6th	5th	4th
☽	4th	3rd	2nd	1st	12th	11th	10th	9th	8th	7th	6th	5th
☉	5th	4th	3rd	2nd	1st	12th	11th	10th	9th	8th	7th	6th
☿	6th	5th	4th	3rd	2nd	1st	12th	11th	10th	9th	8th	7th
♀	7th	6th	5th	4th	3rd	2nd	1st	12th	11th	10th	9th	8th
♂	8th	7th	6th	5th	4th	3rd	2nd	1st	12th	11th	10th	9th
♃	9th	8th	7th	6th	5th	4th	3rd	2nd	1st	12th	11th	10th
♄	10th	9th	8th	7th	6th	5th	4th	3rd	2nd	1st	12th	11th
♄	11th	10th	9th	8th	7th	6th	5th	4th	3rd	2nd	1st	12th
♃	12th	11th	10th	9th	8th	7th	6th	5th	4th	3rd	2nd	1st

Planets are given in the order of the signs they rule - *Publisher's note.*

457. MALE AND FEMALE DEGREES. Many controversies exist as to the sex of the various degrees of the signs, and these differ very much as to their basis. Whatever decrees you elicit from a method founded neither on proof nor analogy nor on the order which the intelligence demands remain obscure until we cease to follow a path

which leads nowhere. There is no sense in people who proceed on such lines, but, nevertheless, they accept indications from the sex of the signs in the same way as from the signs themselves.

Those people, however, who use a method based on order, whatever it may be, do not accept the indications from the sex of signs as a whole, but regard the first degree of a male sign as male, the second as female, the third as male and so on by odd and even, and similarly the first degree of a female sign as female the second as male, etc. as in the case of the male sign. Again there are others who proceed by twelfths of a sign $2\frac{1}{2}^0$ instead of by degrees, just as the whole sphere is divided into twelve signs regarding the first twelfth of a male sign as male, the second as female, and the first of a female sign as female and the second as male, etc., while some of our predecessors considered the first twelve and a half degrees of a male sign to be male, and the second, female, the next two and a half, male and the remaining two and a half, female; proceeding in the inverse manner with the female signs.

With regard to schemes not based on order, a table like that which we append, must be consulted (in which the female degrees are marked with an f).

Aries	7	2f	6	7f	8		
Taurus	7	8f	15				
Gemini	6f	11	6f	4	3f		
Cancer	2	5f	3	2f	11	4f	3
Leo	5	2f	6	10f	7		
Virgo	7f	5	8f	10			
Libra	5	5f	11	7f	2f		
Scorpio	6	7f	4	5f	8		
Sagittarius	2	3f	7	12f	6		
Capricorn	11	8f	11				
Aquarius	5	7f	6	7f	5		
Pisces	10	2f	3	5f	10		

458. BRIGHT AND DARK DEGREES. The distinction drawn between luminous and dark degrees is like the last not founded on any system and consequently recourse must be had to the subjoined table.

Astrologers, however, use it for making decisions as to colours, good and evil, strength and weakness, joy and sorrow, difficulty and ease. But no two books are to be found which agree on this matter, nor are they likely to be found.

The table shows several degrees of light and darkness, brilliant (b), luminous (L), dusky (d), dark or shadowed (s), while some degrees are empty or void (v).

Aries	d 3	s 5	d 8	b 4	s 4	b 5	s 1
Taurus	d 3	L 7	v 2[1]	b 8	v 5	b 3	d 2[2]
Gemini	v 5	b 2	d 3	b 5	v 2	b 6	d 7
Cancer	d 7	b 5	d 2	L 4	s 2	b 8	s 2
Leo	b 7	d 3	s 6	v 5	b 9		
Virgo	d 5	L 4	v 2	b 6	v 4	b 7[3]	v 2
Libra	b 5	d 5	b 8	d 3	b 7	L 2	
Scorpio	d 3	L 5	v 6	L 6	s 2	L 5	d 3
Sagittarius	b 9	d 3	b 7	s 4	d 7		
Capricornus	d 7	b 3	s 5	b 4	d 2	L 4	b 5
Aquarius	s 4	b 5	d 5	b 8	v 3	L 5	
Pisces	d 7	b 4	v 6	L 3	d 10		

459. DEGREES INCREASING OR DIMINISHING FORTUNE. There are also degrees which increase and diminish fortune. The former are those in which if the lord of the period whether sun or moon, or the degree of the ascendant[4] or the part of fortune is situated, the good luck and power of each is doubled. The latter are like pits, in which the planets are enfeebled in their action, being neither able to effect good if lucky nor evil if unlucky - the tendency is therefore towards peace. Both are shown in the following table.

---

[1] Translator has d. [2] Translator has b. [3] Translator has s. *Publisher's notes.*
[4] or Lord of the ascendant.

Degrees increasing fortune in the upper row, pits in the lower.

Aries	19th					
	6th	11th	17th	23rd	29th	
Taurus	8th					
	5th	13th	18th	24th	25th	26th
Gemini	11th					
	2nd*	13th	17th	26th	30th	
Cancer	1st	2nd	3rd	14th	15th	
	12th	17th	23rd	26th	30th	
Leo	5th	7th*	17th			
	6th	13th	15th	22nd	23rd	28th
Virgo	(2nd	12th	20th)			
	8th	13th	16th	21st	25th	
Libra	2nd*	5th	12th*			
	1st	7th	20th	30th		
Scorpio	12th	20th				
	9th	10th	17th	22nd	23rd	27th
Sagittarius	13th	20th	23rd			
	7th	12th	15th	24th	27th	30th
Capricornus	12th	13th*	17th*	20th		
	2nd	7th	17th	22nd	24th	28th
Aquarius	7th	16th	17th	20th		
	1st	12th	14th*	23rd	29th	
Pisces	12th	20th				
	2nd*	9th	24th	27th	28th	

*The * indicates mistakes in MS: brackets ( ) omission.*

460. PLACES INJURIOUS TO THE EYES. There are certain situations which are said to be injurious to the eyes. These have nothing to do with the signs, although some people say that there is a hint of this action in Libra and Scorpius, but they are places which contain certain nebulous stars, or certain animal figures from other constellations which are able to cause this injury. The really nebulous stars are four in number, one in the left hand of Perseus, and this one does not count because its latitude is high, and it is far from the course of the planets; a second, behind the Aselli on the surface of Cancer, this has to be reckoned with; a third is behind the 19th mansion of the moon[1], which is described in books dealing with the heliacal rising of the stars (Kutub al-Anwa) as the venom (Humah) of the Scorpion, and this is of the number, a fourth, as is the tip of the arrow of Sagittarius; again small stars in a group have a cloudy effect such as Haq'ah the 5th mansion of the moon[2] which is composed of three stars in the head of Orion. Ptolemy regarded

[1]The 19th lunar mansion now appears to be the 17th, Al Shaulah, "The Sting".
[2]Al Hak'ah now appears to be the third lunar mansion. See Robson, **Fixed Stars and Constellations in Astrology**. - *Publisher's note.*

them as cloudy, but they need not be included on account of their high latitude. The Pleiades also resemble Haq'ah and belong to this series since their latitude is low, the moon passes by them and the sun also comes near them. Now those two luminaries represent the two eyes and their action vision.

The dangerous places in the animal signs are those like the sting of Scorpius, Nishtar,[1] the (point of the) arrow, Nushaba[2] of Sagittarius, and the Shaukah[3] sharp tail of Capricorn, because its hinder end is fish-like. The hinder end of Leo is also included, as is the star between the eyes of Scorpius and the water below Aquarius, Masabb al-ma. We know of no nebulous star towards the hinder end of Leo except the tuft between his tail and the Great Bear known as Dafirah, which is composed of small stars non-luminous, looking like a cloud shaped like an ivy-leaf, the 'Hulbah' of the Arabs, or tuft of the lion's tail. Its northern latitude is twice as great as the south latitude of Haq'ah, and therefore we think that it cannot be reckoned in this series, besides the dangerous weapons of the lion are his teeth and claws, not his tail. The stars between the eyes of Scorpius extend from the diadem to the heart, and are scattered luminous stars. The water under Aquarius is composed of four small stars near each other situated below the point where the beginning of the flow of water is pictured. Some people call this place the urn of Aquarius, but there are no stars there, and so an urn is assumed in the hand of the man from which the water flows, just as a sword is assumed in the right hand of Perseus.

Our foregoers settled the position of these stars in their time, since which 600 years have elapsed; we however show them in their present position (1340 of the era of Alexander) but it must be remembered that their position increases by a degree every 66 years,[4] i.e. approximately a minute a year.

---

[1] Aculeus. [2] Spiculum. [3] Nashira, or thereabouts. See Robson, **Fixed Stars and Constellations in Astrology**. - *Publisher's notes.*
[4] About 78. The addition of 12° 30' to the above figure gives approximately the present longitude of these stars [1934].

This is the table, and God is all knowing.

Stars from certain signs which harm the eyes

Pleiades	15° ♉ 55' to 17° ♉ 20'
Praesepe	24° ♋
Denbola	5° ♍ to 7° ♍
Between the eyes of Scorpius	15° to 19° ♏
Sting of Scorpius[1]	10° ♐ 40' to 11° ♐ 10'
Venom of Scorpius	14° ♐ 52' to 14° ♐ 54'
Tip of the Arrow	18° ♐ 10' to 18° ♐ 20'
Tail of Capricorn[2]	10° ♒ to 12° ♒
Water of Aquarius	3° ♓ to 4° ♓

We now proceed to consider the conditions in the signs from their relation to the horizon, which we have already referred to as the 'houses' and their adjustment, and we adopt the same order as that used in discussing the indications of the signs and the planets, to facilitate the recognition and comprehension of the data ascertained.

Inshallah ta'aia

---

[1] NB. has here nunir al-fakkah (Alphecca, Corona borealis), present longitude 11°10' Scorpius.

[2] One would assume Shaukah to be a spike of Capricorn, and not the tail, but the longitude corresponds to Deneb al-jadi.

461. SPECIAL INDICATIONS OF THE HOUSES PECULIAR TO NATIVITIES

I   Soul, life, length of life, education, native land.

II   Suckling, nutriment, disaster to eyes if over-taken by ill-luck, livelihood, household requisites, assistants profession of children.

III Brothers, sisters, relations, relations in-law, jewels, friends, migration, short journeys, intelligence, knowledge, expertness in religious law.

IV   Parents, grandparents, descendants, real estate, fields, houses, water supply, knowledge of genealogy, what succeeds death and what happens to the dead.

V   Children, friends, clothes, pleasure, joy, little acquisition of property, accumulated wealth of father, what was said of him at his burial-service.

VI   Sickness, defects of body, overwork, if unfortunate accident to legs, loss of property, disease of internal organs, slaves, maids, cattle.

VII Women, concubines, giving in marriage, marriage-feasts, contentions, partnership, losses, lawsuits.

VIII Death and its causes, murder, poisoning, evil effects of drugs on body, inheritance, wife's property, expenditure, poverty, extreme indigence, feigning death.

IX   Travel, religion, piety, fate, seriousness, attainment of knowledge from the stars and divination, philosophy, surveying, sharp discernment, trustworthiness, interpretation of visions and dreams.

X   Rule of Sultan, government with council of nobles, absolute authority, success in business, commerce, professions, well-behaved children, liberality.

XI   Happiness, friends, enemies, concern for next world, prayer and praise, friendship of women, love, dress, perfume, ornaments, commerce, longevity.

XII Enemies, misery, anxieties, prison, debt, fines, bail, fear, adversity, disease, prenatal fancies of mother, cattle, harbours, slaves, servants, armies, exile, tumults.

462. INDICATIONS RELATING TO HORARY QUESTIONS

I   Asking horary questions, important public matters, nobility, advancement in rank, witchcraft and spells.

II   Examining the querent, lending and borrowing, counting friends, arrival of stranger, enemies or friends, mandate of amir, winds when they blow.

III   Secrets and news and commentaries, well-born ladies, journeys by water.

IV   Old and hidden things, treasures, thieves' hiding-places, schools, fortresses, fetters, [dismissal from office], opening abscesses, lancing and cautery, stepfather, prison.

V   Messengers, right guidance, bribery, rectitude, distant places, poor harvests, securing the wealth of the ancients, feasts, food and drink.

VI   Lost and escaped, some lost trifle which does not turn up, affairs of women and eunuchs, suspicion, hatred, calumny, violence, dissipation, deceit, terrors, prison, enemy, poverty, moving from place to place.

VII   The absent, thief, places where travellers assemble, treasure, death of contemporaries, foreign travel, sudden murder [for a trifle], denial, obstinacy, claiming a right, cheapness and dearness.

VIII   Buried and hidden treasure, things ruined or lost or old, middens and rubbish-heaps, sickness of friends, lawsuits without a case, folly, contention, pride, dullness of the market, leisure.

IX   Failure, abandoned business, books, information, ambassadors, miracles, roads, brothers-in-law.

X   Kings, notables, judges, the celebrated in all classes, amir and his conduct in office, things newly legitimized, wine, stepmother.

XI   The treasury of the Sultan, its officials, trouble in the office, foreigner's child, servants child, things which are sound, beautiful, advantageous, the beginnings of affairs, friendship of the great, bribery, food.

XII   Fugitives, writers, those who neglect devotion, a precious gem, prisoners, the matter which preceded the question, property of oppressors, thieves, lost property, scorn, envy and fraud.

THE HOUSES	463. Their Indications as to Years of Life	464. Opinions of the Hindus as to the Houses	465. Their Indications as to Organs	466. Opinions of the Hindus as to the Organs	467. Rank of Their Powers	468. Their Colors	469. Joys of the Planets in Them	470. Evidence as to the Power of Planets in Them	471. Their Lords According to the Hindus	472. Male and Female
I	Infancy	Soul	Head	Head	12	blue	☿		☿ ♃	male
II	Rest of childhood	Riches	Neck	Face	3	green		♃	☿ ♃	female
III		Brothers	Arm & hand	Arms	5	yellow	☽	♂	☿ ♃	male
IV	Old age Death	Parents Friends	Sides	Heart	7	red		☽	♀ ☽	female
V		Boy ignorant	Heart	Belly	8	white	♀		♀ ☽	male
VI		Enemies Cattle	Belly	Sides	1	black	♂		♀ ☽	female
VII	Prime of of life	Wives	Back Hips	Pelvis	9	mixed	♀		♄	male
VIII		Death	Sexual organs	Sexual organs	4	black		♄	♄	female
IX	Beginning of youth	Journey Debt	Thighs	Thighs	6	white	☉	☿	♄	male
X	Middle	Office	Knees	Knees	11	red		☉	☉ ♂	female
XI	End of youth	Income	Calves	Calves	10	yellow	♃		☉ ♂	male
XII		Expenses	Feet	Feet	2	green	♄		☉ ♂	female

473. CHARACTERISTICS OF THE HOUSES IN GROUPS OF THREE & SIX.

	I II III	IV V VI	VII VIII IX	X XI XII
Body and Soul	Body & Soul: Some say body without soul because it is at a dark place until it emerges into light	Body without soul: Some say body with soul, because it is situated between light and darkness	Neither body nor soul: Because it contains the houses of death and travel	Soul without body: On account of rapid ascension
Right or left	Left	Right	Left	Right
Color	Red	Black	Green	White
Fast or slow	Moderate	Slow	Moderate	Slow
Good or bad luck	Deficient	Good	Deficient	Good
Direction	North	West	South	East
Sex	Female	Male	Female	Male
Temperament	Cold & dry	Cold & wet	Hot & wet	Hot & dry
Hindu ideas as to halves divided by line MC to IMC	From III-I Ascending bow rising, fortunate	From IX-IV Descending bow falling, unfortunate		From X-XII Ascending bow rising, fortunate
Halves divided by line from Hor to Oc.	Nawa = ship Underground Night of planet Allied to rightness & shortness		Chatra = parasol Above ground Day of planet Allied to leftness & length	

474. WHEN HOUSE FORMED OF TWO SIGNS. When a house is formed of two signs, if these are about equally represented, the lords of the signs are also the lords of the house, if both are in aspect; if only one is in aspect it becomes the more important, while if both are inconjunct, that is superior which has the greater number of dignities. The victory must always be given to that one which has the highest number of degrees in the house.

475. PART OF FORTUNE. The Part of Fortune is a point of the zodiac, the distance of which from the degree of the ascendant in the direction of the succession of signs is equal to the distance of the moon from the sun in the opposite direction. The method of determining this is to find the place of the sun (Place 1), then that of the moon (Place 2); the ascendant is Place 3. Then subtract Place 1 from Place 2 beginning with the signs. If in Place 1 this is a higher number add 12 signs to Place 2 and subtract. Next turn to the degrees and subtract as before, if impossible, deduct one sign from Place 2 and add 30° and then subtract. When finished with the degrees, proceed with the minutes the result is the distance of the

moon from the sun. Then add Place 3 by signs, degrees and minutes, and look at the result; if the minutes are more than 59 carry a degree to the degrees, if they are more than 29 carry one to the signs, and if the signs are more than 11, deduct 12, the result is the Part of Fortune.

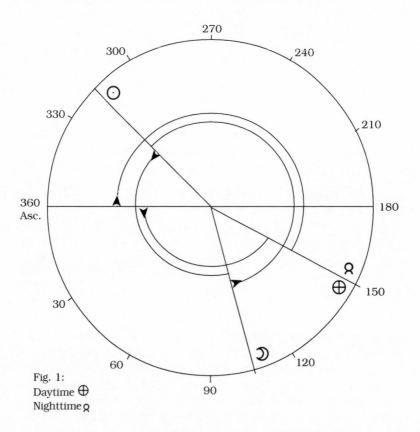

Fig. 1:
Daytime ⊕
Nighttime ♋

The Part of Fortune ⊕ (diurnal) fig. 1 is the distance of ☉ from ☽ according to the succession of the signs measured from the ascendant in the same direction, (nocturnal). Fig. 2 (*next page*) is the distance of ☽ from ☉ according to succession and measured from the ascendant in the same direction. The Part of Daemon ♋ (diurnal) fig. 2 is the distance of ☉ from ☽ according to succession and measured from the ascendant in the opposite direction; (nocturnal) fig. 1 is the distance of ☽ from ☉ according to succession, and measured from the ascendant in the opposite direction. The two inner lines refer to the Part of Fortune; the two outer to the Part of Daemon - modified from BL. fig. 35. [*Translator's note.*]

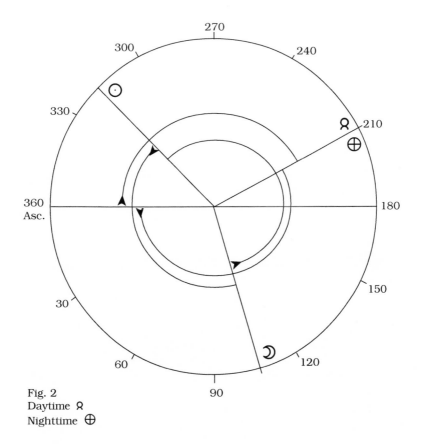

Fig. 2
Daytime ♉
Nighttime ⊕

Take the following as an example. The ascendant is 8°20' of Virgo, the sun is in 27°44' of Cancer, and the moon in 15°25' of Taurus. These are placed in three rows as above described.

	Sun	Moon	Ascendant
	Place 1	*Place 2*	*Place 3*
Signs	03	01	05
Degrees	27	15	08
Minutes	44	25	20

The number of the signs of the sun being higher than that of the moon, 12 must be added, making 13, from which the 3 of the sun must be deducted, leaving 10. The degrees of the sun are also higher than those of the moon, therefore 1 must be deducted from the signs, leaving 9, and 30 added to the degrees, making 45, from which 27 falls to be subtracted, leaving 18. Similarly with the minutes 1 degree must be carried to them, leaving 17 and 60 added making 85 from which 44 subtracted leaves 41. The result of the subtraction of the sun's place from the moon's is therefore 9s 17° 41', to which the place of the ascendant being added gives 14s 25° 61'. From the last figure 60 must be deducted and carried as 1 degree to the degrees, and from the first 12 must be deducted leaving 2, so that the result, the Path of Fortune, is 2s 26⁰ 01', viz. 26° 01' of Gemini.

This is the method of calculation adopted by Ptolemy for the Part of Fortune which he never altered, but others proceed in this way for diurnal nativities while for nocturnal ones they put the moon[1] in the first place, the sun in the second, and the ascendant in the third, whence necessarily many disputes.

476. OTHER LOTS THAN PART OF FORTUNE. Ptolemy recognized only one Part of Fortune, but others have introduced an excessive number of methods of casting lots at nativities. We reproduce in tables those which Abu Ma'shar has mentioned.

In each case there are three things to be attended to, Place 1 The beginning, Place 2 The end, and Place 3 The casting-off point, which are treated as in the preceding paragraph, the position in a figure of the heavens of the fortune or lot in question being thereby determined. These three points are called respectively, 'manqud' 'manqud minhu' and 'muzad'alaihi'.[2] Sometimes the same arrangement is used for both diurnal and nocturnal nativities, but frequently points 1 and 2 are interchanged for nocturnal ones.

It is impossible to enumerate the lots which have been invented for the solution of horary questions, and for answering enquiries as to prosperous outcome or auspicious time for action; they increase in number every day, but the following 97 different lots, 7 of which belong to the planets, 80 to the houses and 10 to neither are those most commonly in use.

---

[1] In which case the ⊕ would be in ♏ 20° 39' at the same distance from the ascendant in the direction of succession.

[2] The amount subtracted, that from which it is subtracted, the amount added.

Numbers	Names of the Fortunes	*Distance between Place 1 & Place 2*		*Cast from Place 3*	*Diurnal or nocturnal*
	Fortunes of the Seven Planets				
1. ⊕	Part of Fortune or Lunar horoscope	☉	☽	Asc	Change
2. ☊	Part of Daemon and religion	☽	☉	Asc	Change
3.	Of friendship and love	⊕	☊	Asc	Change
4.	Of despair & penury & fraud	☊	⊕	Asc	Change
5.	Of captivity, prisons and escape therefrom	♄	⊕	Asc	Change
6.	Of victory, triumph and aid	☊	♃	Asc	Change
7.	Of valour & bravery	♂	⊕	Asc	Change
	Fortunes of the Twelve Houses *First House - Three Fortunes*				
8.	Of Life	♃	♄	Asc	Change
9.	Pillar of horoscope Nativities, Permanence Constancy	⊕	☊	Asc	Change
10.	Reasoning & eloquence	☿	♂	Asc	Change
	Second House - Three Fortunes				
11.	Property	Lord of II	Cusp of II	Asc	Change
12.	Debt	♄	☿	Asc	Change
13.	Treasure trove	☿	♀	Asc	Same
	Third House Three Fortunes				
14.	Brothers	♄	♃	Asc	Same
15.	Number of brothers	☿	♄	Asc	Same
16.	Death of brothers & sisters	☉	10^0 of III	Asc	Change
	Fourth House - Eight Fortunes				
17.	Parents	☉ (♃)	♄	Asc	Change
18.	Death of Parents	♄	♃	Asc	Change
19.	Grandparents	II	♄	Asc	Change
20.	Ancestors & relations	♄	♂	Asc	Change
21.	Real estate a/c Hermes	☿	♃	Asc	Change
22.	Real estate a/c some Persians	♄	☽	Asc	Change
23.	Agriculture, tillage	♀	♄	Asc	Same
24.	Issue of affairs	♄	Lord of ♂ or ☍	Asc	Same
	Fifth House - Five Fortunes				
25.	Children	♃ (♀)	♄	Asc	Change
26.	Time and number of sexes	♂	♃	Asc	Same
27.	Condition of males	♂	♃	Asc	Same
28.	Condition of females	☽	♀	Asc	Same
29.	As to whether expected birth male or female	☽	Lord of the house of ☽	Asc	Change

Al Biruni said that an illiterate soothsayer's accurate prophecy was due to the coincidence of his ☊ with his ascendant. [*Translators's note.*]

Numbers	Names of the Fortunes	*Distance between Place 1 & Place 2*		*Cast from Place 3*	*Diurnal or nocturnal*
	Sixth House - Four Fortunes				
30.	Disease, defects, time of onset of them a/c Hermes	♄	♂	Asc	Change
31.	Disease a/c to some of the ancients	☿	♂	Asc	Same
32.	Captivity	Lord of time	Lord of house of lord of time of lord of VI		Same
33.	Slaves	☿	☽	Asc	Same
	Seventh House Sixteen Fortunes				
34.	Marriage of men (Hermes)	♄	♀	Asc	Same
35.	Marriage a/c Walis	☉	♀	Asc	Same
36.	Trickery and deception of men and women	☉	♀	Asc	Same
37.	Intercourse	☉	♀	Asc	Same
38.	Marriage of women (Hermes)	♀	♄	Asc	Same
39.	Marriage of women (Valens)	☽	♂	Asc	Same
40.	Misconduct by women	☽	♂	Asc	Same
41.	Trickery & deceit of men by women	☽	♂	Asc	Same
42.	Intercourse	☽	♂	Asc	Same
43.	Unchastity of women	☽	♂	Asc	Same
44.	Chastity of women	☽	♀	Asc	Same
45.	Marriage of men & women (Hermes)	♀	Cusp VII	Asc	Same
46.	Time of marriage (Hermes)	☉	☽	Asc	Same
47.	Fraudulent marriage & facilitating it	♄	♀	Asc	Same
48.	Sons-in-law	♄	♀	Asc	Change
49.	Lawsuits	♂	♃	Asc	Change
	Eighth House - Five Fortunes				
50.	Death	☽	Cusp VIII	Degree of ♄	Same
51.	The Anairetai	Lord of Asc	☽	Asc	Change
52.	Year to be feared at birth for death, famine	♄	Lord of house in which ♂ or ☍	Asc	Same

Numbers	Names of the Fortunes	Distance between Place 1 & Place 2		Cast from Place 3	Diurnal or nocturnal
	Eighth House - Five Fortunes (continued)				
53.	Place of murder and sickness	♄	♂	Degree of ☿	Change
54.	Danger and Violence	♄	☿	Asc	Change
	Ninth House - Seven Fortunes				
55.	Journeys	Lord IX	Cusp IX	Asc	Same
56.	By water	♄	15° ♋	Asc	Change
57.	Timidity and hiding	☽	☿	Asc	Change
58.	Deep reflection	♄	☽	Asc	Change
59.	Understanding and wisdom	♄	☉	Asc	Change
60.	Traditions, knowledge of affairs	☉	♃	Asc	Change
61.	Knowledge whether true or false	☿	☽	Asc	Same
	Tenth House - Twelve Fortunes				
62.	Noble births	From lord of time to his degree of exaltation		Asc	Change
63.	Kings and Sultans	♂	☽	Asc	Change
64.	Administrators, vazirs, etc.	☿	♂	Asc	Change
65.	Sultans victory conquest	☉	♄	Asc	Change
66.	Of those who rise in station	♄	⊕	Asc	Change
67.	Celebrated persons of rank	♄	☉	Asc	Same
68.	Armies and police	♂	♄	Asc	Change
69.	Sultan. Those concerned in nativities	♄	☽	Asc	Same
70.	Merchants and their work	☿	♀	Asc	Change
71.	Buying and selling	☊	⊕	Asc	Change
72.	Operations and orders in medical treatment	☉	♃	Asc	Change
73.	Mothers	♀	☽	Asc	Change
	Eleventh House - Eleven Fortunes				
74.	Glory	⊕	☊	Asc	Change
75.	Friendship and enmity	⊕	☊	Asc	Change
76.	Known by men and revered,	⊕	☉	Asc	Change
	constant in affairs			Asc	Change
77.	Success	⊕	♃	Asc	Change
78.	Worldliness	⊕	♀	Asc	Change
79.	Hope	♃	☿	Asc	Change
80.	Friends	☽	☿	Asc	Same
81.	Violence	☊	☿	Asc	Same
82.	Abundance in house	☽	☉	Asc	Same
83.	Liberty of person	☿	☉	Asc	Change
84.	Praise & acceptation	♃	♀	Asc	Change

Numbers	Names of the Fortunes	Distance between Place 1 & Place 2		Cast from Place 3	Diurnal or nocturnal
		Twelfth House - Three Fortunes			
85.	Enmity a/c to some of the Ancients	♄	♂	Asc	Same
86.	Enmity a/c Hermes	Lord XII	Cusp XII	Asc	Same
87.	Bad luck	☊	⊕	Asc	Same

Altogether Seven Fortunes belong to the Planets and 80 to the houses.

Ten Fortunes not related to Planets or Houses.

88.	Hailaj	Degree ♂ or ♂°	☽	Asc	Same
89.	Debilitated bodies	⊕	♂	Asc	Change
90.	Horsemanship, bravery	♄	☽	Asc	Change
91.	Boldness, violence, and murder	Lord Asc	☽	Asc	Change
				Asc	Change
92.	Trickery and deceit	☿	☊	Asc	Change
93.	Necessity and wish	♄	♂	Asc	Same
94.	Requirements and necessities a/c Egyptians	♂	Cusp III	Asc	Same
95.	Realization of needs & desires	⊕	☿	Asc	Same
96.	Retribution	♂	☉	Asc	Change
97.	Rectitude	☿	♂	Asc	Change

477. DIFFERENCES IN PRACTICE. There are people who adopt methods differing from the above under some circumstances; e.g. with regard to the lot of parents when Saturn is under the rays of the sun, they take from Jupiter to the sun by night or vice versa by day, and cast from the ascendant. Again in the case of the lot for grandparents, if the sun is in Leo, they take from the beginning of Leo to Saturn by day, and by night in the opposite direction. And if it is in the domicile of Saturn then from the Sun to Saturn by day, and vice versa by night, in both cases cast from the ascendant even if Saturn is under the rays or otherwise afflicted. Should two lots indicate the same point, it is regarded as very fortunate. In some of these cases the instructions for day and night are the same, in others different as may be seen from the table, in the former event there is no advantage to be derived from a separate calculation.

478. OTHER LOTS. This matter of casting lots is a very long one, so that one might think there is no end to it. For instance there are those which are cast at the turn of the year (the entry of the sun into Aries) on worldly matters and affairs of empire, and those which are cast at conjunctions and oppositions of the moon to elicit prognostics as to weather, as to success of ventures, and other horary

questions. We append in tables the opinions of others on these matters which we have derived from books on the subject.

479. LOTS WHICH ARE CAST AT ANNIVERSARIES OF THE WORLD-YEAR, AND AT CONJUNCTIONS.

Numbers	Names of the Fortunes	Distance between Place 1 & Place 2		Cast from Place 3	Diurnal or nocturnal
1.	The sultan's lot	MC ☉	MC anniv.	♃	Same
2.	By another way	Deg. Asc. Conj.	Deg. conj.	Asc	Same
3.	Victory	☉	Lord of VII (deg. of Dsc.)	Asc	Same
4.	Battle	♂	☽	Deg. Lot of Victory	Same
5.	Second way a/c to Umar	♂	☽	Asc	Same
6.	Third way al-furkhan	♄	☽	Asc	Same
7.	Truce between armies	☽	☿	Asc	Same
8.	Conquest	☉	♂	Asc	Same
9.	Triumph	⊕	♃	Asc	Change
10.	Of 1st conjunction	Asc. year conj.	Degr. conj.	Asc	Same
11.	Of 2nd conjunction	Asc conj.	Degr. conj.	Asc	Same

The following lots are associated with the years, the four quarters, and the conjunctions and oppositions of the moon.

1.	Earth	♄	♃	Asc	Same
2.	Water	☽	♀	Asc	Same
3.	Air and wind	☿	Lord of his domicile	Asc	Same
4.	Fire	☉	♂	Asc	Same
5.	Clouds	♂	♄	Asc	Change
6.	Rains	☽	♀	Asc	Change
7.	Cold	☿	♄	Asc	Change
8.	Floods	☉	♄	☽	At moon rise

Num-bers	Names of the Fortunes	Distance between Place 1 & Place 2		Cast from Place 3	Diurnal or nocturnal
	Lots as to prognostics regarding crops &c.				
1.	Wheat	☉	♃	Asc	Change
2.	Barley, meat	☽	♃	Asc	Change
3.	Rice, millet	♃	♀	Asc	Change
4.	Maize	♃	♄	Asc	Change
5.	Pulse	♀	☿	Asc	Change
6.	Lentils and iron	♂	♄	Asc	Change
7.	Beans, onions	♄	♂	Asc	Change
8.	Chick-peas	♀	☉	Asc	Change
9,	Sesame, grapes	♄	♀	Asc	Change
10.	Sugar	♀	☿	Asc	Change
11.	Honey	☽	☉	Asc	Change
12.	Oil	♂	☽	Asc	Change
13.	Nuts, flax	♂	♀	Asc	Change
14.	Olives	☿	☽	Asc	Change
15.	Apricots	♄	♂	Asc	Change
16.	Water melons	♃	☿	Asc	Change
17.	Salt	☽	♂	Asc	Change
18.	Sweets	☉	♀	Asc	Change
19.	Astringents	☿	♄	Asc	Change
20.	Pungent things	♂	♄	Asc	Change
21.	Raw silk, cotton	☿	♀	Asc	Change
22.	Purgatives	☿	♄	Asc	Change
23.	Bitter purgatives	♄	♂	Asc	Change
24.	Acid purgatives	♄	♃	Asc	Change

Num-bers	Names of the Fortunes	Distance between Place 1 & Place 2		Cast from Place 3	Diurnal or nocturnal
	Lots cast in connection with horary questions				
1.	Secrets	Lord of Asc	Cusp X	Asc	Same
2.	Urgent wish	Lord hour	Lord Asc	Asc	Change
3.	Time of attainment	Lord hour	Lord X	Asc	Change
4.	Information true or not	☿	☽	Asc	Change
5.	Injury to business	Lord Asc	⊕	Asc	Same
6.	Freedmen and servants	♃	♄	☿	Same
7.	Lords and masters	♃	♄	☽	Same
8.	Marriage	♀	Cusp VII	Asc	Same
9.	Time for action (Walis)	☉	♃	Asc	Same
10.	Time occupied therein	☉	♄	Asc	Same
11.	Dismissal or resignation	☉	♃	♄	Same
12.	Time thereof (Walis)	Lord of the affair	⊕	Cusp X	Same

Num-bers	Names of the Fortunes	Distance between Place 1 & Place 2		Cast from Place 3	Diurnal or nocturnal
Lots cast in connection with horary questions, *continued*					
13.	Life or death of absent person	☽	♂	Asc	Same
14.	Lost animal	☉	♂	Asc	Same
15.	Lawsuit	♂	☿	Asc	Same
16.	Successful issue	☉	♃	Asc	Same
17.	Decapitation	☽	♂	Cusp VIII	Same
18.	Torture	☽	♄	Cusp IX	Same

480. ARROWS AND ANIMALS. On the practice of sortilege by two arrows and the interpretation of animal omens.

The book of Hermes known as the **85 Chapters** discusses the indications derived from both. As to omens from two animals, Masha'allah mentions that a black animal should be interpreted as Saturn and a yellow one as the Sun. As to sortilege by two arrows none of the interpreters has been helpful in furnishing an explanation with regard to them except Masha'allah whose examples are founded on the lives of kings. Other members of the profession are inclined to adopt long calculations by many and devious methods neither restricted nor free from error. Some of them at the entrance of the Sun into Aries in discussing the permanence of empire and the probability of rebellion take the first arrow as equivalent to the distance of the sun from the middle of Leo, and the second to that from the moon to the middle of Cancer, both cast from the ascendant, and the same for day and night, while others who have studied the subject most earnestly assert that the first arrow represents Saturn himself and the second Jupiter. What has been written on this subject alone would make two large books.

481. CAZIMI, ORIENTALITY, OCCIDENTALITY. We now proceed to deal with the various positions of the planets in relation to the sun, which are responsible for the most complete changes which closely resemble changes in their indications, due to the vicissitudes of natural conditions.

If a planet should be within less than 16' of conjunction with the sun or have passed it by less than the same amount, it is designated as 'samim' [cazimi]. The superior planets, however, are only in such a position in the middle of their direct course, while the inferior planets are in it in the middle of their direct and retrograde courses. In regard to orientality, the inferior planets in the middle of their retrograde course resemble the superior in the middle of their direct course. If the superior planets and the inferior ones in the middle of the retrograde course exceed the minutes of tasmim all are said to be 'muhtariq', combust, until their distance from the Sun is 6°; thereafter they are no longer so styled but are said to be under the rays. In this condition they remain like prisoners in confinement until the distance of Venus and Mercury from the sun amounts to 12⁰, of Saturn and Jupiter to 15° and of Mars to 18°. This point is described as the beginning of 'tashriq' orientality,* but they are not necessarily visible at this period, for the time of visibility varies with each country and climate. But the term tashriq is properly limited (to the heliacal rising) and after this they are designated 'musharriq', which the Persians call 'kanar-i ruzi'. Thereafter the higher planets differ from the lower, for the former continue tending eastward till they are 30⁰ from the sun, and after which they are said to be weakly oriental till a distance of 90° is attained, and the name tashriq does not cease to be applied for at sunrise they are in the eastern quarter, while whenever the 90° is exceeded the term orientality ceases to be applicable. Thereafter the first stationary point is reached, after which the retrograde movement sets in; when this is concluded there is again a stationary point before the direct course is entered. Arrival at opposition to the sun occurs in the middle of the retrograde path, which is thus divided into two sections, lst and 2nd.

The higher planets after their station until they are distant 90° from the sun are in the east at sunset, but when less than 90° incline to the west, and when the distance is 30° this situation is called the beginning of occidentality (taghrib), till Mars is 18°, Saturn and Jupiter 15°, and thereafter they are under the rays, until only 6° separate them, when they are combust until only 16' remain when they are again in tasmim.

In the **Almagest** the opposition of the higher planets to the sun is called the beginning of the night, it is a situation which is peculiar to the higher planets, for under it they rise at sunset. The Persians however, are in the habit of using the expression Kanar-i shab for both higher and lower planets, but that condition which they call the beginning of the night is really occidentality, therefore they add west, so as to distinguish between the two.

---

*They are now west (right) of the sun, rise before it in the east, and become morning stars.

74

482. INFERIOR PLANETS AFTER TASHRIQ. We said that the orientality of Venus and Mercury occurs on the retrograde path[1] and is not completed till a distance of 30^0 from the sun in both cases. Thereafter they are stationary and then comes the direct course to their greatest (western) elongation, after which they again begin to approach the sun. All of these situations are called oriental, until 12^0 separate them from the sun, the beginning of their matutine occultation in the east. They are under the rays until 7^0 from the sun and are then combust till they reach the limit of samim and conjoin with the sun in the middle of their direct course.[2] Thereafter they pass out from samim, when their situation in the west resembles that of the higher planets in the east to the extent which has been noted of them in regard to combustion and being under the rays and visibility at evening twilight. Then they gain their greatest eastern elongation and stop before they again retrograde, passing through all the stages the distances of which we have noticed till they return to tasmim on the retrograde course.

483. HOW VENUS DIFFERS FROM MERCURY HERE. It is necessary to distinguish between Venus and Mercury as regards orientality and occidentality, as has been done between Mars on the one hand and Saturn and Jupiter on the other, (astronomers are agreed that no such distinction is necessary between these two planets) for Venus has a very high latitude, and sometimes conjunction occurs when it has attained its highest north latitude,[3] it then remains visible, so that the expressions combust and under the rays cease to be applicable, although the planet is in those positions; similarly at tasmim when the north latitude exceeds $7°$, it must not be described as samimah nor muhtariqah but simply as accompanying the sun, muqarinah.

484. RELATIVE POSITION SUN AND MOON. The position of the moon with regard to the sun as to tasmim and combustion is similar to that of the other planets, as long as the distance is less than 7^0 east or west of the sun; beyond that it is under the rays till the distance increases up to $12°$ which is approximately new moon; thereafter the various distances described as phases which produce the quarter, half, three quarters and complete illumination succeed, and are followed at the same distances on the other side of opposition by similar figures.

485. POSITION RIGHT AND LEFT OF SUN. Astronomers agree that all three higher planets from the time of conjunction to opposition, and both lower planets from conjunction on the retrograde to

---

[1] After inferior conjunction. [2] Superior conjunction. [3] 6^0 22'.

that on the direct course, and the moon from opposition to conjunction are to the right (west) of the sun, while the higher planets from opposition to conjunction, and the lower from conjunction on the direct to that on the retrograde course, and the moon from conjunction to opposition are on the left (east) of the sun.

486. INFLUENCE OF PLANETS UNDER CHANGED CONDITIONS. It may be asked whether with the changes in situation of the planets described, their action also changes. If their action did not change, there would be no advantage in paying attention to these situations. Astrologers are however agreed that the maximum influence of the planets is at tasmim, and during this the indications are of happiness and good news; they are also agreed that such influence is at its minimum in combustion, until it arrives at a point where unluckiness changes to ruination. However, distinctions are made in accordance with the concord and discord of the nature of the planets, as e.g. heat may become increased and moisture diminished, consequently the injurious influence of combustion is less with some planets and greater with others. After conjunction, the planet, when under the rays, is like a sick person advancing to convalescence, and when oriental attains full strength and is in a position to bestow all its benefits. The Persians call this its vazirate,* (and any one who wishes to do a good act, does it at this time.) They extend this name to the whole of the position right of the sun, until at a distance of 30° from the sun the beneficial action begins to stop, and the indications of happiness to become moderate, till at 60° the action changes, this point is called the minor unlucky point, 75° the middle unlucky point, and combustion (on the retrograde) the major unlucky point. The planet at the first resting place appears strangled, hopeless, in the first section of the retrograde course sluggish and depressed, while in the second section hope of succor is given, which is confirmed in the second station, delivery being near at hand, while the direct course indicates, as its name suggests, prosperity and power. Similarly the nature of the planets alters from their rising to their setting in the eccentric orbit, being dry during the former and moist in the latter, without however the nature of their action being affected. Also from rising to setting in the orbit of the epicycle, for from the oriental phase to the first stop they are moist, then to the middle of the retrograde course warm, then to the second stop, dry, and back to orientality cold. The reason of the change in the orbit of the epicycle is that the action of the latter is bound up with the sun, and it is said that nearness to the sun means dryness and distance moisture. Combustion also changes the nature and other conditions like rising and setting which

---

* Position of authority.

bring about action in the epicycle different from that in the eccentric orbit. The circumstance that the planet is posited in moist places of the signs or terms gives friendliness; again, in the matter of maleness and femaleness they change, becoming male when oriental and female when occidental.

Again among the signs the planets also are affected by the indications of the whole sign, just as the soul depends on the condition of the body, and so a male planet becomes effeminate when in a female sign, and is even affected by the male and female degrees of a sign, so that there are mixed indications of eunuchism and hermaphroditism, effeminate men and masculine women.

So also in quadrants of the sphere in relation to the horizon the planets may change in the matter of sex, and also at the cardines. The effect of situation at the cardines however is simply to increase the influence of the planet, so that good fortune at a cardinal point is increased, especially if the sign be a fixed one. Calamity and adversity are also intensified in a fixed sign especially if cadent to the cardines, while they are weakened in a tropical sign especially if not cadent.

Some people assert that the west is favourable to the lower planets, and the east to the higher, but you must understand that this is derived only from the analogy of maleness and femaleness, the east being male and the west female, while the criterion of the difference between them is distance from the sun.

It has been shown that the orientality of the superior planets occurs on the direct course after combustion, on this account they are then more powerful because as it were, they are escaping from distress and calamity; comparable to this is the vespertine visibility of the inferior planets, which also concurs after combustion on the direct course.

The occidentality of the superior planets occurs likewise on their direct course as they proceed to combustion, so to this is comparable the matutine occultation of the inferior planets also on their direct course. The orientality of the inferior planets resembles that of the superior ones in as much as in both cases it takes place after combustion: if the inferior planets were at that time direct there would be entire agreement of all in the matter of orientality. But the occidentality of the inferior planets, when their movement becomes slow, is a much more injurious and weakening influence than the occidentality of the superior ones because the former have now turned their faces both towards the retrograde course and combustion; so the superior planets in their occidental phase are safer than the inferior, because it is only succeeded by their occultation.

We have extracted from Ya'qub b. Ishaq' al-Kindi* all that a beginner requires to know with regard to the different indications of the planets as to their powerful influence in orientality and their weakness in occidentality, although these differences do not amount to being exact opposites.

---

* The "Philosopher of the Arabs" - 9th Century. For his philosophical work cf. Hugel, Al-Kindi, Leipzig 1857: for his scientific work Wiedemann, "XXVI, XLII, XLIV: for his astrological writings Loth, Al-Kindi ale Astrolog, Leipzig 1875.

	487. INDICATIONS WHILE ORIENTAL	488. INDICATIONS WHILE OCCIDENTAL
♄	Beginning of old age, happy in farming and art of irrigation, profound and effective judgment, sharp and authoritative dispatch of all business matters.	Advanced old age, miserable standard of living, business mean and small in extent, work in connection with irrigation and wells, poor food, fraud.
♃	Beginning of manhood and maturity, good conduct, beauty, elegance, desirous of office as vizir or qadi so as to insure justice, many possessions, good reputation, joy in children.	Advanced middle age, occupations of moderate importance, position as prefect or law-agent, and all things connected with religion such as copying books of traditions; immoral acts, pilgrimage, sufficient wealth.
♂	Leading in battle, commanding armies, reputation for courage eagerness for conquest; quickness in business; success in mining.	Mean positions in the army such as butcher, cook, smith, farrier, surgeon; theft; work to do with fire and iron.
☉	Tashriq and taghrib indicating position relative to sun are inapplicable to the sun itself.	
♀	Actions when oriental are less effective than when occidental.	Beauty, hatred, love, joy, gladness, pleasure, marriage, gifts; as to crafts, forbidden pleasures, work with colours, pictures, brocades, embroidery.
☿	Intelligence, reasoning power, long consideration, wise decisions, poetry, eloquence, clerk of taxes, surveyor, orderliness, affability, medicine, astrology.	Same as under tashriq but less efficient; occidentality occasions little harm to it and to Venus.
☽	From middle of month to 22nd denotes mature manhood, thereafter to conjunction, old age.	From conjunction to 7th day, childhood, from there to opposition, youth; when the moon is under the rays it points to things secret and concealed, and especially it points to the ill condition of creatures resembling the light at that stage.

489. APPLICATION AND SEPARATION. The terms application and separation refer to the formation of aspects between the planets and withdrawal from such positions. These are dependent on the signs, and the same names are employed which we have already explained in regard to the aspects of the signs, viz. conjunction, two sextile, two square, two trine and opposition. When two constellations are in aspect, planets within them are also in aspect, when the former are not in aspect, the planets in them are inconjunct and concealed from each other. When two planets are in the same sign or in two signs in aspect to each other and at the same degree, they are said to be conjoint in reality, and the one whose orbit is lower is said to apply itself to that whose orbit is higher, because the lower one is swifter and constantly overtakes the slower one. Consequently the moon applies itself to all the planets and is applied to by none; Mercury applies itself to all except the moon, Venus to all except the moon and Mercury, the sun Mars and Jupiter to those above them, Saturn alone applies itself to no planet because all are below it. When of two planets in aspect, the degrees of the inferior one are less than those of the superior planet, the inferior one is said to be proceeding to conjunction and when greater to be separating from the superior. At the time of conjunction the lower planet is said to be conferring counsel (dafi' tadbir) on the higher and the latter receiving counsel from it. This is conjunction in longitude.

490. BEGINNING OF APPLICATION. As application is like meeting, and separation like parting, so an inferior planet when it enters a sign where it comes into aspect with a superior one, begins to show its movement towards conjunction, which increases till conjunction is completed, unless something else intervenes such as its being outstripped by another planet, or deserted by the superior planet leaving the sign in which it was before completion, or by itself becoming retrograde and thus frustrating completion. But there is much difference of opinion as to the amount and limits of completion. Some people say that it begins at 5 degrees and continues till the degrees are equal, the 5 'dead' degrees, being made the basis of this interpretation. Others say 6 degrees, because it is the fifth of a sign, and the average of the planets' terms. Others again say 12 degrees, the distance at which the light of the moon is obscured by the sun, and still others, 15 degrees, the orb of the sun, while others say the average of the respective orbs of the planets in question. Again many assert that only complete conjunction need be attended to.

Separation begins when the degree of the inferior becomes even a minute higher than that of the superior, but, on account of the trace of influence which remains, the completion of separation should be determined by the amount assigned to the beginning of application.

491. THE 'DEAD' DEGREES. The 'dead' degrees referred to are five degrees beyond the ascendant in the direction opposite to the succession of the signs. Ptolemy does not reckon these as belonging to the twelve houses, and does not regard them as cadent to the ascendant, but if there is a planet in them he associates it with the ascendant.

492. CONJUNCTION ONLY IN LONGITUDE? There are two other kinds of conjunction besides that in longitude, viz. in latitude and in nature. The former occurs when the latitude of two planets is the same either north or south, and the degrees of latitude are equal. Then they are said to be conjoint by latitude. If the degrees are not equal one must look whether that of lower latitude is rising in the quarter in question and whether that of higher latitude is setting in the same quarter, if so, they are said to be moving towards conjunction. If the latitude of the setting planet is lower than that of the rising one, they are separating. If both are rising one must see whether the extreme latitude of the lower is not less than that of the higher one, if so, they are moving towards conjunction; if less, that cannot occur. If both are setting, and that which has the higher latitude is quicker in setting, they are said to be proceeding to conjunction, whether that is completed or not, because that of lower latitude may move to the other side (or the other may overtake it).

The superiority of conjunction by latitude to that by longitude is due to the fact that it does not occur except when the planets are in aspect.

There is another advantage viz. that suppose an inferior planet applies itself to a superior in longitude and then to a third in latitude which is inconjunct to the superior one, then the latter does not continue in conjunction by longitude at the same time.

Conjunction by nature occurs when two planets are in equipollent signs, and takes place when they arrive at corresponding degrees in these. E.g. Jupiter is in 20^0 of Aries and the moon in 5^0 of Pisces, when the latter has attained to 10^0 of Pisces, which is the corresponding degree to 20^0 of Aries, the conjunction by nature is completed. The condition becomes fortified if the planets are in aspect during this conjunction in nature. Similarly if two planets are in corresponding degrees in signs correspondent by direction, the conjunction is complete; e.g. Jupiter is at 20^0 of Aries and the moon at 5^0 of Virgo, then complete conjunction will occur at 10^0 of Virgo. Aspect here also improves the condition.

493. TESTIMONY AND DIGNITY. The expressions testimony and dignity are synonymous terms and are applicable to a planet in two different ways. One concerns the fortunate position which it may occupy, if e.g. it should be lord of the house in which it is situated, or be in its exaltation, or in any other position congenial to it, it may have one or more of these dignities. If however it is not in a favourable situation it is said to be peregrine, while if either in its detriment, or its fall, calamity is added to the alien situation.

The other kind results from something outside the situation of the planet, and is of three varieties. First, when it is in a situation favourable to another planet and on this account has the advantages of that other attributed to it, whether that be lordship of a house or exaltation; Second, depending on the disposition or essential nature of the planet, as e.g. the testimony of Mars is connected with war and lawsuits, of Jupiter with riches and estates, of Venus with amusement and marriage; Third, dependent on time, such as day for the sun and night for the moon, or the lordship of the day or hour or the like.

494. ORDER OF PRECEDENCE OF DIGNITIES. The dignities have a certain order of precedence. Most important is the lordship of the house, next, exaltation, then, term, then triplicity, lastly, face; and so a certain scale of numbers has been assigned to them, viz. 5 to the house, 4 to exaltation, 3 to term, 2 to triplicity and 1 to face. The dignities of the various planets may therefore be added up and compared with each other, so as to see which is preeminent. It is related that an authority on this subject assigns 30 to the lordship of the ascendant, 20 to exaltation, 10 to lordship of face, 5 to that of term, 3½ to that of triplicity, 4½ to that of the hour, and finally to the sun or moon, whichever is lord of time, as much as to the lordship of the ascendant. The figures are then added and compared. This is the practice of the Astrologers of Babylon and Persia, who regard the lordship of the face as very important. But among the Astrologers of the present day, the triplicity is regarded as having precedence over term and face, and indeed the latter is often considered of no account. Further in certain circumstances changes may take place in this order, e.g. the lord of exaltation may take precedence over the lord of the house in matters of empire and government in high places. It is necessary to know that these dignities are strengthened by aspect, or by other conditions which re-

place aspect, because if the numbers of two planets add up equal, one of which is in aspect and the other inconjunct, the former is preferred even if its favourable positions and testimonies amount to less than those of the latter.

495. THE RULING PLANET "AL-MUTEN". The word 'mubtazz' means a victorious planet and victory may be arrived at in two ways: 1. mutlaq absolute (essential), dependent on dignities due to position in the orbit, or in relation to other planets or to the horizon; 2. muqayyad limited (accidental), when these dignities are referred to one of the characteristic properties of the twelve houses.

496. HAYIZ AND HALB. The terms 'hayyiz' and 'halb' are related in meaning, and they share one condition viz. that when a diurnal planet is above ground by day and beneath it at night, and when a nocturnal planet is above ground at night and beneath it by day, it is said to be in its halb, and a planet is described as in or not in its halb. When in addition to this a planet is male and in a male sign or female and in a female sign, the condition is called hayyiz, and a planet is said to be in or not in its hayyiz. Moreover it is obvious that hayyiz is more comprehensive than halb, because every hayyiz is a halb but not every halb a hayyiz.

Abu Ma'shar in this matter has increased the number of male and female degrees. It should be known that Mars in this matter of hayyiz is different from the other planets, because it is both male and nocturnal; if it is above the earth by night and below it by day and in a male sign, it is then in its hayyiz.

497. CONTENTION. Munakara (contention) is nearly the reverse of hayyiz and occurs when a diurnal planet is in the domicile of a nocturnal one, and the latter is in the domicile of a diurnal planet; or when a nocturnal planet is in the domicile of a diurnal one, and the latter is in the domicile of a nocturnal planet.*

498. JOY OF THE PLANET. The planets are said to be joyful, powerful, happy and in good spirits when they are in congenial sections of the signs, in their halb or hayyiz; the quarters friendly to them N. S. E. or W., and also when far from the sun those which were previously in distress, like the superior planets when oriental and the inferior when occidental in their direct course. They are also in their joys in those preferred houses which we discussed - this is the best known of all - and finally they are joyous in those quadrants depending on the horizon; the superior in the increasing quadrants

* E.g. ♄ in ♈ and ♂ in ♓ or ♂ in ♓ and ♃ in ♉.

the inferior in the decreasing ones.

499. PROSPERITY AND ADVERSITY. Prosperity is associated with the cardines, as these indicate a happy mean; adversity with the cadent houses, which point to destruction and excess. Being in those houses which are succedent to the angles is beyond the half-way line to prosperity, for they are the paths leading there from adversity. But this prosperity and adversity are not all alike, just as the cardines are not alike but are higher and lower in glory and dignity. And indeed the cadent houses are not alike in their destructive influences, because although the 3rd and 9th houses are cadent, the 6th and 12th are not only cadent but are also inconjunct to the horoscope.

500. BESIEGED. A planet is said to be besieged when situated between two others, as e.g. when a planet in sign 1 is surrounded by others in signs 2 and 12. It also occurs when three planets are in one sign, the middle one whose degree is less than the one and higher than the other is said to be corporally besieged. Again a planet is described as besieged by the rays, when in front of it is another in sextile or quartile and another behind it in like aspect. When besieged by two infortunes the influences are extremely bad, while if between two fortunes, they are extremely good.

501. SUSPICION. A planet on which a number of unfavourable conditions is heaped, and in evil case on account of being combust or retrograde, or in its detriment or fall or in a cadent house, or inconjunct, or antagonized by infortunes, or whose aspects are inimical is said to be suspect in its significance. (If it shows any promise, it is unable to carry it out.)

502. BENEFACTION AND REQUITAL. If a planet is in its fall or in a pit or in a sign in which it has no proper section, it is as it were confined in a tight place or cave. If now one of the planets friendly to it or its sponsor applies itself to it, and gives a helping hand to deliver it from its calamitous situation, it is described as conferring a favour on it, and is called a benefactor. Again if the benefactor arrives at a similar disadvantageous situation, and is applied to by the first planet, this is called requital.

503. TWO RIGHT AND TWO LEFT HANDS. When a planet is in the cardo of mid-heaven and its sextile and quartile rays fall together above earth, it is said to have two right hands, if these fall below earth two left hands. The indications of the former are success and victory.

504. VOID OF COURSE. If while within a sign a planet does not enter into conjunction with another, although in aspect to other planets, it is said to be void of course, and is regarded as having separated from conjunction whether in that sign or not. (This name is given to it because the field is empty and it moves without any companion.)

505. FERAL. When a planet is in a sign and no other planet has been in aspect with it from the time of its entry to that of its exit, it is said to be feral in its course. This is practically impossible with the superior planets and the sun, and can rarely occur, but with the moon it is necessarily the case and frequently occurs. If the moon with its rapid motion did not exist, this might very well happen to the inferior planets, one of them moving rapidly the other slowly. Some people say that when the moon is feral, this is a substitute for conjunction with the planets in whose terms it happens to be within the sign, but this opinion is trivial and quite unsupported.

506. COMPLETION OF CONJUNCTION. That a 'familiarity'* of the various forms discussed should be completed between the inferior planet conferring counsel and the superior receiving it, there must neither be return nor evasion nor intervention nor refranation nor abscission of light nor prevention. Each of these will be distinguished and interpreted.

1. Return. This happens to a superior planet when retrograde or under the rays, for from weakness it is unable to hold what is offered to it, therefore returns and does not accept it. If the situation is such that there is reception between them, or if the inferior planet is at an angle or both of them are at angles, or succedent houses, the end of such return is satisfactory. If however the inferior planet is in the weak situation described, and the superior one at an angle or in a succedent house, the result is destructive even if at first hope was indicated. If both are in a weak situation from the beginning to the end there is nothing but destruction and ruin.

2. Evasion. This occurs when an inferior planet is about to conjoin with a superior one; before this takes place the latter moves out of the sign, and the inferior planet applies itself to another planet either in the same or another sign, the first aspect never having been completed.

---

* An astrological expression for relation of planets by aspect, conjunction, equality of declination (antiscions) &c.

3. Intervention. This occurs when an inferior planet tends to become conjunct with a superior one, in the latter part of whose sign there is posited a third planet lower than the superior and higher than the inferior planet. Before the inferior planet completes conjunction, the third intermediate planet retrogrades towards the superior planet and passes it by, till the inferior inevitably conjoins with it and not with the superior one.

4. Abscission of light. If it should happen that the intermediate planet is not in the same sign as the superior one, but in the next and retrogrades into it, this intervention is one of two methods of cutting off the light. The second method is when an inferior planet tends to conjoin with a higher one and a third still higher is situated towards the latter part of the sign, then before the inferior planet can conjoin with the intermediate one, the latter moves to the higher one and passes or conjoins with it. The inferior planet does not conjoin with the intermediate one but with the superior later.

5. Refranation. If an inferior planet tends to conjunction with a superior one but before completion becomes retrograde the familiarity is said to be frustrated by refranation.

6. Prevention. When there is a third planet in a sign between the inferior and superior ones, it prevents the conjunction of the former with the latter until it itself has entered into conjunction. When two planets tend to form a familiarity with a third at the same time, the one by means of corporal conjunction, the other by aspect, the former renders the latter vain if their degrees are equal; but when their degrees are different, and the one casting the aspect is nearer to completion than the one tending to conjunction, the former is preferred. (When however two planets apply themselves by aspect to a third at the same time, that is preferable, whose reception occurs first.) Certain aspects must have an advantage over others, just as corporal conjunction has over aspect, so that the more powerful aspect should interfere with the weaker, but astrologers have not pronounced on this matter.

507. RECEPTION. When an inferior planet arrives in one of the dignities proper to a superior one, and makes known to it the relation thus established, there is an exchange of compliments such as 'your servant' or 'neighbour'. If further the superior planet happens to be in a situation proper to the inferior one, mutual reception takes place, and this is fortified, the richer the situation is in dignities, especially when the aspects indicate no enmity nor malevolence. When reception does not take place the result is negative.

508. THE DONOR. We have already stated, 489-506, that al-dafi' is the application of one planet to another and is described as bi'l-tadbir. The inferior planet making application is not specified as dafi' unless it is in a dignified situation proper to it without regard to the situation of the superior, the madfu'ilaihi, this conjunction is called daf al-quwwah, and if in one proper to the superior planet, daf al-tabi'ah, which is the same as qabul described above, or else the inferior planet is in a place proper to itself which also happens to be congenial to the superior planet; this is called daf al-tabi'atain, because the natural properties of both are united. The same expression is used when one (an inferior) planet in its hayyiz conjoins with another (a superior one) in its hayyiz, the planets being both either diurnal or nocturnal, for the hayyiz requires two conditions to render it complete.

509. FOLLOWER. When a retrograde inferior planet follows and overtakes a retrograde superior one, the situation is called 'muradafah'. Here there is no return on account of the similarity of their situations but if there is reception the indication is for the successful termination of business which was threatened with ruin. However this conjunction, although there is no refusal, is not equal to one in the direct course but is far inferior in significance.

510. SUBSTITUTES FOR CONJUNCTION AND ASPECT. There are other conditions which are efficacious besides aspect and conjunction. When an inferior planet and an intermediate one both apply themselves to a superior planet, the latter is called a collector 'jam', because it assembles the light of the others. If these are in aspect to each other, this is just as good as conjunction with the collector; if they are not in aspect, that collection of their light by another planet is effective in place of conjunction although they are inconjunct.

If an inferior planet separates from an intermediate one which is inconjunct to a superior planet, and thereafter conjoins with the superior one the light of the intermediate planet is transferred to the latter. This is called 'naql' or translation and occurs between two planets which are (inconjunct or between two which are) in aspect but far from conjunction. This condition is effective in lieu of conjunction.

There is another form of translation, when the inferior planet conjoins with the intermediate one, and the latter has already been in conjunction with the superior planet; it is just the same as if the inferior had applied itself to the superior. This occurs when the inferior planet is inconjunct to the superior one because, when in aspect, it is swift in arriving at the conjunction with the superior one.

In the books, one always finds the naql of Mars by the sun to Saturn described as the greater naql and that of the moon by the sun to Saturn as the lesser naql. If two planets are inconjunct to a third or to a certain place in the zodiac, and then both conjoin with one which is in aspect to both and also to that third or that place, the result is like mirrors reflecting from house to house. This has also been called 'radd' but in view of what we have said before about radd, the use of that word is ambiguous. There is also another aspect of real translation which is not much enlarged upon, except in relation to separation; they say that when an inferior planet withdraws from conjunction with a superior one and conjoins with another then naql occurs, light being transferred from the one to the other and as this translation is an effective substitute for conjunction, it follows that it should not be void of the effect of separation. However some other word than 'radd' should be used for this condition perhaps sarf or 'aks (conversion or inversion) to remove the ambiguity.

511. OPENING THE DOORS. When two planets whose natures are opposed conjoin, this is called opening the doors. So the conjunction of the sun or moon with Saturn indicates quiet rain, fine drizzle or snowstorms, that of Venus and Mars torrential rain, hail, thunder and lightning, and that of Mercury and Jupiter the opening of the doors of the winds.*

512. STRENGTH AND WEAKNESS OF THE PLANETS. In dealing previously with the relations of the planets to the sun, to each other, to their own orbits, to the zodiac, and to the horizon, we have discussed as far as possible the good and evil effects of each, as well as the summed effects of more than one. Each planet has a most favourable situation, and when some advantage is lost, its power is diminished to a like extent. The converse is true with regard to unfavourable situations.

A planet is at the height of its power when the following conditions are present. Motion direct, rapid and increasing, far from the sun's rays, oriental if superior, occidental if inferior, in aspect to both sun and moon, and these in a fortunate state, besieged by fortunes or aspecting them, relieved of infortunes, associated with fixed stars of the same character, rising in its own orbit, passing above the infortunes and below the fortunes, north latitude increasing, happening to be in domiciles of the fortunes, or their

*If you see the moon separate from Venus and apply itself to Mars or v.v., this is also opening of the doors: Haly, p. 396. [Based on sign rulerships.]

huzuz or in a place resembling its own nature, or in houses most congenial to it, in its own hayyiz, at an angle or succedent thereto, in a quadrant of the same nature, and increasing, elevated high above the malefics and conquering.

But when slow, [retrograde, under the rays, occidental if superior, and if inferior moving slowly] westward towards retrograde, inconjunct to sun and moon, or in an unfriendly aspect to them, without reception, the infortunes in an inimical aspect, or besieged by them, associated with fixed stars of a contrary nature, setting in own orbit, so that the malefics pass above and the benefics below, decreasing south latitude, in unlucky houses, in parts of signs foreign to them, in detriment or fall, in a contrary hayyiz, distant from the angles or succedent houses, in a quadrant of different nature, at the nadir of their joys, and conquered by the malefics high above them; this is the acme of weakness. But in all conditions there is always an admixture of good and bad, often difficult to interpret, and requiring all the resources of the art as well as experience and industry.

513. HOW SUN AND MOON DIFFER FROM THE PLANETS IN THIS REGARD. In regard to the foregoing there is considerable difference between the sun and moon on the one hand and the other planets on the other. When both of the luminaries are in aspect to each other, and to the benefics, and are in their own sections of the signs or those of the benefics, both of them are strong. But if they are in situations unsuitable to them, and the malefics, full of enmity are above them, and the benefics below, or are eclipsed, or near the dragon's head or tail, especially the latter by less than 12⁰, both of them are weak. The moon is especially so when near or in conjunction, or on the wane, or under the earth, or in the combust way, all of which increase its weakness.

Many people include among the inauspicious situations for the moon the being in the last part of a sign, and in the 12ths of both malefics, setting in the south, and being in the ninth house from the ascendant, all of which are not exclusively applicable to the moon, especially the last part of the signs, where all the terms belong to the malefics, a situation bad for all the planets, as are the 12ths referred to. There is also to be considered the quarter of the heaven, and the fact that the ninth house from the ascendant is the nadir of the joy of the moon, and this is peculiar to it.

514. THE COMBUST WAY. The combust way is the last part of Libra and the first of Scorpius. These two signs are not congenial to the sun and moon on account of the obscurity and ill-luck con-

nected with them, and because each of them is the fall of one of the luminaries. They also contain the two malefics, the one by exaltation (Libra, Saturn), the other by house (Scorpius, Mars). The peculiarity however which has given the name muhtariq is that the exaltation of Saturn is near, the fall of the sun being on the one hand and that of the moon on the other, while the adjacent parts of both signs are occupied by terms of Mars.

515. DIVISIONS OF JUDICIAL ASTROLOGY. There are as many divisions of Astrology as there are elements in the universe. These may be either simple or compound and on both the influence of the planets is active. The former on the whole do not submit to such influence, nor to any change, except where they come into contact with each other, when, because they are mutually opposed and violent, they are always in strife. Such admixture does take place on the surface of the earth, but is only completed by the heat of the sun's rays. So all four elements become united, and the surface is the place appointed for the action of the planets, which extends as far as the power of their rays penetrates by reason of the presence of interstices. Then these rays return by a contrary motion and carry with them the aqueous vapour which they have produced, and they rise from the earth until they reach a point where the power of such movement becomes weak. So this motion and agitation is the cause of all the vicissitudes and disasters of nature, the resultant phenomena being either permanent or temporary.

Anything therefore in the way of heat or cold or moderate temperature, of moisture or dryness owing to movements of the atmosphere, or of the various forms of moisture carried by the winds such as cloud, rain, snow; everything that is heard in the air such as sharp claps and rolls of thunder; everything that is seen such as lightning, thunderbolts, rainbows, halos, meteors, also shooting stars, comets and similar atmospheric phenomena; everything that occurs in the earth in the way of tremors, and subsidences, and in the water as tempests and floods, and the flux and reflux of the tides - all these form the subject matter of the first division of astrology. These phenomena are not permanent or rarely so; rain, snow, comets and earthquakes are those which have the longest duration; were they not sufficiently widespread their concentration in one spot would be disastrous.

A second division is that which is concerned with the mixed elements, such as occur in plants and animals, and is of two kinds, affecting the whole of a population or only a part thereof. Famine may be taken as an example of the former, due to failure of crops or drought, and epidemics such as spread from country to country, like the plague and other pestilences which depopulate cities.

The latter variety is more localized and scattered in its appearances, it results from psychical phenomena, such as battles, struggle for power, change of dominion from one land to another, deposition of kings, revolutions, emergence of new religions and sects, so that this chapter is a long one and this variety the more important of the two.

The third division is specially concerned with the environment of the individual human or other, the events which affect him in the course of his life, and the influences which remain behind him and in his progeny, while the fourth has to do with human activities and occupations. All of these are founded on beginnings or origins possibly trivial.

Beyond these there is a fifth division where such origins are entirely unknown. Here astrology reaches a point which threatens to transgress its proper limits, where problems are submitted which it is impossible to solve for the most part, and where the matter leaves the solid basis of universals for one of particulars. When this boundary is passed, where the astrologer is on one side and the sorcerer on the other, you enter a field of omens and divinations which has nothing to do with astrology although the stars may be referred to in connection with them.

516. PRINCIPLES BY WHICH INQUIRIES BELONGING TO THE FIRST DIVISION ARE KNOWN. The fundamental principles which are applicable to enquiries in the first and second of these divisions of astrology are substantially the same. They are based on the greater, intermediate and lesser conjunctions, the exact places at which these occur and the ascendants at these times; further on the thousands known as hazarat, hundreds, tens and the firdaria. There are people who take from the conjunction and opposition of the moon which preceded the enquiry, and substitute this for the above, and there are others who depend on the nearest eclipses past or future, of which the most hurtful are those of the sun, especially if of considerable extent.

517. ANALYSIS AND INTERPRETATION OF THESE. The degrees at which Saturn and Jupiter meet in conjunction, together with the ascendant of that time, and the ascendant of the year of the conjunction all move in the direction of the succession of signs through a whole sign in a complete solar year. The point arrived at is called

a terminus; moreover, this terminus of each year is in the sign next after that in which it was the year before, and in the same degree thereof, e.g. if the terminus of the first year was in 10° of Cancer, that of next year would be in 10° of Leo. The matter of the thousands and what follows them is in the like case, and there is no difference between them except in the different amount of time allotted to the degrees and signs. This is a usage of the Persians and became known to us through their language.*

We have stated before that according to Abu Ma'shar the years of the universe are 360,000, the deluge being in the middle of these. This statement occurs in his book called '**The Book of Thousands**' where the degrees of the zodiac are each made equal to a thousand years, so that the fraction belonging to a year is 3 3/5 seconds. This is the great division; secondly, the signs are made equal to a thousand years each; this is the term of thousands. Thirdly the signs are made equal to single years, the terminus of years being thus produced as we said before. Fourthly the degrees are made equal to single years, and this is the small division.

Between the units and thousands two other terms are introduced, one in which each sign equals a hundred years and another in which each is ten. Nothing is said with regard to the share of the degrees in the case of the tens and hundreds such as we have spoken of in the case of the thousands and units.

We have previously discussed the extent of the firdaria, and placed in a table their order at nativities. But here the order changes and that of the signs which contain the exaltations of the planets is adopted; viz. 1st Aries which has the exaltation of the Sun; 2nd Taurus of the Moon; 3rd Gemini of the Dragon's Head; 4th Cancer, of Jupiter; 5th Virgo, of Mercury; 6th Libra, of Saturn; 7th Sagittarius, of the Dragon's Tail; 8th Capricorn, of Mars; 9th Pisces, of Venus. The order is therefore, Sun, Moon, Dragon's Head, Jupiter, Mercury, Saturn, Dragon's Tail, Mars, Venus, and then back to the Sun. The distribution of partnerships is as before, but the lords of exaltation have precedence over the lords of the firdaria, which however preserve their own order and the partnership in their own sec-

---

* According to Abu Ma'shar in his Kitab when the heavens were first set in motion all the planets, the sun included, were in conjunction; when the same phenomenon again presents itself, which may not occur for millions of years, the world will enter on a new period. Reinaud, Abu'l-fida, I. CXCI. **The Book of the Thousands** on religious houses treats of birth, duration and end of the world, and fixes the times when great changes in Empires and Religions will take place. d'Herbelot, IV. 695.

tions, except in the case of the Dragon's Head and Tail, which do not enter into partnership and are therefore alone in their firdaria.

These are the principles which must be relied upon and used at every anniversary of the world-year* and its quarters, also at every conjunction and opposition of the moon, especially those which occur immediately before the anniversary and the quarters.

518. REVOLUTIONS REFERRED TO AT CONJUNCTIONS AND THEIR QUARTERS. The revolutions which are mentioned in connection with conjunctions have a duration of 360 solar years. They are divided differently into quarters, by some people equally into 90 years each as if quarters of the ecliptic, by others, substituting the relative duration of the seasons of the solar year into a first quarter of 90 years, a second of 85¼, a third of 90 and a fourth of 94 3/4.

519. PRINCIPLES SPECIAL TO THE SECOND DIVISION AND DIFFERENT FROM THOSE OF THE FIRST. In addition to the principles laid down for dealing with questions of the first order, the following are adopted for those of the second. The turn of the solar year and of its quarters, the conjunctions, oppositions, quarters and other phases of the moon, also the experiences of people in all places as to the rains on the days of the past year, further, the eclipses, combustions, conjunctions, retrograde movements of the planets which have occurred in the year. There are astrologers who note the ascendant at the time of the entry of the sun and moon into the signs, and deal in the same manner with the five planets, but this is obviously going out of the way without advantage.

520. LORD OF THE YEAR. That planet is known as the Salkhuda (Persian for) lord of the year which, at the anniversaries of the world-year (solar year), is at the ascendant or one of the angles with dignities in its own degree, or if there is nothing there, that which is in a succedent house. If there is nothing there also, then it is that planet which is not inconjunct with the ascendant or its lord. According to the Hindus it is that planet which is next in order of the lords of the days; to each planet a year being given. They make a great deal of this.

521. PRINCIPLES OF THE THIRD DIVISION. The principles adopted for questions of the third order are as follows:

For every creature there is a time of its first appearance,

---

*The entrance of the Sun into Aries. But in 1020 the Perigee must have been some 14° E of the winter solstice in which case the relative duration of the Seasons would be Summer 92.8, Spring 91.4, Winter 88.6, Autumn 88.12.

and decrees are then sought from the ascendant and the figure of the heavens as to its condition. This section is exclusively devoted to man, and must not be employed for plants, crops or animals. There are two initial points, sowing or conception, and time of appearance or birth. From the arrangement of the stars, the haylaj becomes known, and the kadkhuda[1] the ruling planets of the houses, the gifts (allowances of length of life), the additions, and deductions therefrom, and the murderers which put an end to it.

At the anniversaries of the birth there become known the progressions, the apheses,[2] the lord of the revolutionary figure, the divisor or distributor of the fortunes of life, and the mudabbir its partner in administration, the lords of the weeks, and the firdaria.

522. INTERPRETATION IN DETAIL. As to the analysis and interpretation of these, the infant is at first delicate and weak, is unfavourably affected by the least change in its condition, and it is impossible to have confidence in its survival until it has attained the age of four years. These are called the years of rearing by the astrologers. The first thing they do in these years is to ascertain whether it is going to survive or not, and when in their opinion it is sufficiently strong to be reared, they look whether there is a haylaj or not. This they search for in five places; 1. The lord of the time, day or night; 2. The moon by day and the sun by night; 3. The degree of the ascendant; 4. The Part of Fortune; 5. The degree of conjunction or opposition of the moon preceding the birth. The haylaj is one of these. After it has been determined according to the proper rules,[2] then the most powerful planet as regards dignities of those in aspect to it is the kadkhuda. If it is at an angle a large number is assigned, if succedent an intermediate one, and if in a cadent position a small one. These are the numbers which we discussed under the years of the planets and according to the condition of the kadkhuda as regards power or weakness, these numbers indicate years of life or months or days or hours.

These are the gifts or allowances of the kadkhuda. In the event of its being in a maleficent or weak position, every fortune which is in a friendly aspect to it, or is in reception with it, adds its smallest number to the allowance, in the form of years or months according to the strength or weakness aforesaid, while every infortune in inimical aspect deducts such a number. These are styled the additions and deductions. The result is the longest period of

---

[1] Persian for head of the household. [2] Plural of aphetic.
[3] The haylaj must be in an aphetical place, either near the East or West Angles or in the IX, X or XI house.

life to which the native can attain, if one of the anaeretai[1] does not interfere. Sometimes in a nativity there is no haylaj, in which case the length of life must be estimated from the numbers of the fortunes present. The anaeretai are moreover malefic in themselves and their rays are inimical like certain fixed stars which are known for their evil effects. When the direction arrives at them, at the time when the half-year or quarter or yearly allowance is due, disaster results and then the fortunes can do nothing against the unfavourable situation.

There are astrologers who regard the situations at the thirds of the year as gifts of the kadkhuda in place of the positions at the quarters. But there are many anaeretai, among them the degrees of the ascendant and of the moon, if one of them interferes with the other, and again the cusps of the 4th, 7th and 8th houses. These are separately dealt with in the books.[2]

Each year the ascendant is ascertained when the sun comes round to the same minute of the ecliptic in which it stood at the birth, i.e. the anniversary, and also every month when the sun arrives at the same degree and minute it occupied in the radical or revolutionary figure. The lord of the ascendant at birth is the lord of the first year, that of the second, the planet next below in the order of the spheres, and so the lords of the revolutionary figures for succeeding years are reached in the same fashion as you proceed with the lords of the hours. The Babylonians adopt the same method, but start with the lord of the hour of birth, instead of that of the ascendant, the second being next in order below.

The termini of the years are determined as follows: a sign being given to each year, the end of the second year is in the second sign at the same degree as the ascendant[3] and so with the third. When the signs and degrees of the yearly terms have been learnt, each year is divided into (thirteen) months of 28 days 1 hour 51 minutes and a sign to each given, so that the last month ends at the same degree as the radical ascendant has the same sign as the first, while the first month of the next year has the same sign as the

[1] Saturn and Mars. [2] Apparently Capella was regarded as one. [Fixed star at 22° II (2000). - *Publisher*] When Abu Sahl on leaving Khwarizm with Avicenna was overtaken by a sandstorm he foretold his death within two days because the direction of the degree of his Ascendant would then reach Capella (not Capricorn as in translation), Chahar Maqala p. 87.
[3] According to Hermes, **De revol. nativ. II** p.219 and Junct. p.1051 the dominus anni is the lord of the sign of the year (as distinct from the Salkhuda of world-years), and to Wilson p.280 that planet which has most dignities and is strongest in a revolutionary figure.

year; similarly a sign is given to each of thirteen periods of 2 days 3 hours 50 minutes, the end of the last of these periods coinciding with the end of the monthly term.[1]

The lord of the week is determined as follows: take the days elapsed since birth and divide by 7, note the product, and count on the same number of signs from the ascendant of the radix, the one you arrive at is the sign of the week.[2] Then count the remainder which is less than 7 from the lord of the ascendant in the direction opposite to the succession of the signs, the sign you thus arrive at is the lord of the day of the week in question.[3] There are astrologers who proceed in the direction of the signs, not contrary thereto.

523. OTHER THINGS TO BE RECKONED WITH. We have referred previously to the Apheta and its direction in regard to termini, the thousands and cycles. Here its meaning requires to a certain extent to be explained, because in nativities the aphesis is not calculated by the equal degrees of the ecliptic but by degrees of ascension. So the aphesis from the degree of the ascendant and the planet which is situated there is calculated by oblique ascension at the locality in question, one year for each degree.[4] So also the aphesis of the planet at the occident angle will be according to its descension at the locality, because the setting of any sign at a locality is equal to the ascension of its nadir. However with regard to the M.C. and I.C. and any planet situated there, the aphesis is in all localities by ascension in the right sphere. So if a planet is not transiting one of these four degrees but a point between two angles, its ascension is compounded of those of the adjoining angles, and the calculation is a long and difficult business.

An arc of direction is always calculated from the haylaj, the significator of life, and never from any other point except in special cases. The Kadkhuda is the significator for the length of life. The degree of the ascendant is always made apheta whether there is a haylaj or not. When at an anniversary or any other time there is

---

[1] Cf. Junctinus p.1138 who is more accurate. The year is divided into 13 months of 28d 2h 17m 3s, 9''', 14'''', and the month into 30 days of 22h 28m 35s 16''' 18'''' !

[2] Of the last complete week. [3] But probably not its real lord.

[4] This is Ptolemy's method of determining the length of life by the time taken by one planet to reach a certain point of the zodiac or the former position of another planet by the diurnal movement calculated in planetary hours (1/12th of its diurnal arc) or degrees of oblique ascension. A year being assigned to each degree, 90 years would be the allowance if the points were separated by the semi-diurnal arc, which converted into degrees of right ascension might be considerably more.

ascertained the point at which the direction of the haylaj has arrived, the lord of the term in question is called qasim or divisor,* in Persian, jan-bakhtar, bringer of the fortunes of life. The name qasim comes from the circumstance that because life is situated between the radical place of the haylaj and the anaeretic point, the interval is divided into sections by the terms of the signs, and the lords of the terms become the lords of these sections. Any planet which is in the term of the apheta or directs its rays to it becomes associated with the administration of that section.

With regard to the ruling planets: in the various houses of the planets are numerous dignities and associated therewith pre-eminence in the possession of these. The mubtazz without qualification is that planet which at a nativity is predominant by virtue of numerous dignities at the ascendant or its lord, or at the five aphetic points in the radix and similarly at its anniversaries. The firdaria we have already discussed both in relation to the years of the world and to nativities.

524. PROCEDURE AT A NATIVITY. Procedure to be observed at a birth.

When the child is born you must take the altitude of the sun if it is day, and work out the ascendant and its degree. This is the horoscope of the nativity. If it is night, then the altitude of a well-known fixed star which is on the rete of the astrolabe must be taken. Do not concern yourself with the planets which would only involve you in difficulties, nor with the moon, for working with it would be a mistake unless it is necessary. Further if by day or night the condition of the heavens is such by reason of cloud or the like, that you cannot get an observation, then only the determination of the time remains.

When you know how much of the day or night has passed, the ascendant can be calculated by the method we have described. The number of hours elapsed can be determined in two ways, the first by having a water-clock or other apparatus for measuring time going before the labour comes on, the clock having been set by sunrise or sunset or the like. When the birth takes place, the hour must be noted. The other way is to set the clock going at the time of birth if previous notice has not been received, and watch it until it is possible to take the altitude of the sun or a star. It is then pos-

*The divisor is important for indicating the profession a native should enter. Junct. p.1070 from Albohazen Haly f. 95 and also to a certain extent 'alcelcadeny', 520. p. 255, see VI. 3.

sible by counting back the numbers of hours shown by the clock to get the exact time.

If there is no clock available, all that is necessary is a cup of any material which will hold water; a hole must be made in the bottom of any dimension you please, and when the child is born you may proceed in one of two ways at choice, first by letting water into it and second by allowing water to escape from it. If you choose the former, place the cup on the surface of clean water, watch till it fills and sinks. Immediately take it out and empty it, and place on the water again, and count the number of times it sinks until the sun or a star is visible. A mark must then be made at the point the water has reached, to indicate the fraction to which it had sunk. Then take the altitude and note the time, and proceed as before till as many sinkings, together with the fraction marked, have taken place as noted. Then take the altitude again and determine the number of hours from the second time the cup was placed on the water, and count back the same amount from the time the sun became visible, which gives the time of birth.

If you choose the second way, place the cup on something like a trivet, and take a pitcher full of water, and fill the cup, when all the water has poured or trickled out, fill again and count the numbers of pitchers used till the sun or star is visible, if there is water in the cup make a mark, and proceed as before with the determination of the time.

525. IF TIME NOT NOTED USE OF 'ANIMODAR'. Should no observation have been made at the time of birth, the determination of that time is beyond the reach of science, for there is no way of knowing it, but astrologers by estimation and conjecture arrive at one little different in the sign of the ascendant, when an attentive observer employs cautious questioning. But it is necessary that there should be a certain degree for the ascendant, so they find a way, by using an indicator (namudar) which furnishes one which they assume to be the degree desired. The indicator most in use is that of Ptolemy, which if it does not disclose the exact degree, is the best substitute. The method in question is to ascertain as accurately as possible the time communicated to you, and determine the ascendant, the cardines and the places of the seven planets. Then find the degree of the conjunction of the moon which occurred before the birth if that was in the first half of the month, or else the degree of opposition, if in the latter half. Then determine which planet has the most dignities and testimonies, then the one that comes next, and so proceed with the others till the last and note the

result. The most important testimony is being in aspect to that degree, for when two planets are equal in the number of their dignities, the one in aspect whatever that may be, is preferable. Then examine which of the two most dignified planets is nearest to an angle by counting the number of their degrees. Thereafter transfer the angle to the degree of the nearest planet and derive the ascendant from that. If the degrees of the two planets are very distant from an angle, take the next planet in order of dignity, and examine the others till you find that which is nearest to an angle and proceed as before.

There are astrologers who do not attach any importance to the relative distance from or nearness to an angle but simply make the degree of the angle which is nearest to the most dignified planets the place (from which to derive the ascendant) without altering its degree to that of the planet and proceed as we have said.

526. TIME OF CONCEPTION. The essential condition which makes it possible to discover the temperament, constitution and form of a native as well as the conditions which take place in him during life within the mother's womb is the ascertainment of the time of conception. Authorities insist of use being made of this. It is possible to learn from the mother or the father if they agree the beginning of the phenomena of pregnancy, the direction of which they have month by month or week by week ascribed to Saturn or Jupiter and so down through the spheres.

The procedure adopted by astrologers is founded on two principles either of which is satisfactory if properly executed: 1. It is assumed that the degree of the ascendant at birth is the same as the degree at which the moon stood at the time of conception, and 2. Conversely, that the degree of the ascendant at the time of conception is the same as that in which the moon stood at the time of birth. In the first place it is desirable to ascertain from the mother whether it is the 7th, 8th, 9th or 10th month of pregnancy, having done so look at the ascendant and the configuration of the heavens at the time which has been approximately arrived at; if the moon is at the degree of the ascendant, give to the ascendant of conception the same degree. Then the child has completed so many full revolutions of the moon before birth, either 7 (191 days 6 hours), 8 (218 days 13 hours) - here be careful not to say that an 8 months child is not viable - 9 (245 days 21 hours) or 10 (273 days 5 hours).

If the moon is not at the degree of the ascendant, whether above or below the earth, if above, look how many degrees separate them, and take a day for every 13^0 11' * - and for every degree 1 hour and 5/6, and every minute of a degree 1 5/6 minutes of time, and subtract the result in days hours and minutes from the days of that month of which you have been informed. If the moon is below the earth, take the distance from the ascendant to the moon, and proceed in the same way, but add the result to the days of the month in question. So the greater or less number which you arrive at is the time spent by the infant in the womb. Count back therefore from the time of birth the number of days and hours, the result is the time of conception. Thereafter ascertain the position of the moon, and when you know its degree make this the degree of the ascendant, for this is approximately accurate from the data available.

527. THE FOURTH DIVISION. For horary questions of the 4th order, the ascendant of the beginning of the matter in hand must be ascertained, whether that be determined already as in the case of a nativity, and therefore known, or whether a time has to be selected or chosen as a starting point. The purpose of this section is to select a suitable time for carrying out some business so as to insure the presence of fortunes and the absence of infortunes, just as we protect ourselves on the surface of the earth from the rays of the sun, by selecting northern aspects, and shady spots and using moistened punkahs and icehouses. In this matter pay no attention to the silly talk in which the Hashwiyites persist and their denial of what we have accepted in this matter of 'elections'.

The essence of this section is so to adjust the cardines that the malefics are as distant as possible both in themselves and their rays, while they are to be kept illuminated by the benefice and their light, especially the ascendant and its lord, also the moon and the lord of its house, and the significator of the business which is the subject of the inquiry. Also see to the moon and the lord of the ascendant and the significator that they are in aspect to each other, and place them in such a position that they all cast an aspect to the ascendant lest the election should turn out to have bad effects. This is a long and wide field of enquiry into which it is impossible now to penetrate further.

528. THE FIFTH DIVISION AND ITS PRINCIPLES. Rules for questions of the fifth order.

In view of the fact that the nativities of querents regarding various contingencies are for the most part unknown, astrologers deal with the statement of the querent as a starting point just as if

* The mean tropical movement of the moon in a day.

it were a nativity. The ascendant of the time is taken and investigated, as well as its lord and the moon and that planet which the moon is leaving. These are used as significators for the querent, and as the matters on which guidance is sought belong for the most part to the 7th house and its lord, or to such other house in which the question is comprised and its lord, also to that planet with which the moon is about to conjoin, there is no reason why with a little care and attention an answer should not be found somewhere among the twelve houses. This division is known as that of the questions.

529. IDLE AND GENERAL QUESTIONS. In case of an idle request or one for a general prognostic the custom of the majority of astrologers is to follow the same procedure as in other questions, namely to ascertain the ascendant of the time of the query. They then examine the aspects as they would at a nativity and make conclusions i.e. as to the remaining period of life and the conditions therein.

There are however astrologers who increase the range of horoscope inspection by claiming to elicit the past life of the querent. Hashwiyite astrologers, inclined to falsification, when such a question is asked bid their clients return and sleep on the matter for three nights and concentrate their attention on it during the day, and then question them. After satisfying myself as to their writings I know of no method of dealing with them except insisting on exposing their vicious decrees and their leading the querent into crime by the bad advice given him.

530. THOUGHT READING. Khabi' refers to hidden objects (concealed in the hand) and damair to secret thoughts reserved by the querent. What greater ignominy is likely to be the part of Astrologers than that resulting from hasty dealing with such questions and in comparison how numerous are the lucky hits of Magicians who keep up a patter while they are on the lookout for telltale indications and actions!

Now we have arrived at a point of the science of the stars which I have regarded as sufficing for the beginner; any one who exceeds the limits set out above exposes himself and the science to derision and scorn, for such are ignorant of the further relations of the art and especially of those which have been ascertained with certainty.

Conclusion of the **Book of Instruction on the Elements of the Science of Astrology** Composed by Abu al-Raihan Muhammad b. Ahmad al-Biruni. May the Mercy of God be upon him. Abundant Mercy. And His blessings on Muhammad, his descendants the pure in heart.

All glory be to God first and last.

As the Colophon has no date, the following from the first fly-leaf of the MS are added.

By the accident of time this book came into the possession of the poor dependent on Allah the all-sufficient Aubad b. As'ad b. Mihrla'r al-Mustaufi. May the Most High God improve his circumstances, and favour the realization of his hopes in this world and the next. May he cause him to select aright the winning arrow from the quiver. In the month of Allah, Rajab the Deaf, 839 AH. (Jan. 1436 A.D.)

He, the Guide. This book came into the possession of the poor slave in need of the Mercy of our Lord the Creator 'Ala b. Al-Hunain b. 'Ala al-Sahqi. May God overlook his sins by Muhammad and his family and his generous associates. In the year 889 AH. (1485 A.D.) Praise be to God first and last and may He bless our Lord and prophet Muhammad, the best of mortals, and all prophets and saints.

Well endowed is he who with sufficient humility unites intellect and Soul

For these two form a fortunate star-conjunction which has an enduring influence with the people.

INDEX

104

Better books make better astrologers.
Here are some of our other titles:

Derek Appleby
Horary Astrology: The Art of Astrological Divination

C.E.O. Carter
An Encyclopaedia of Psychological Astrology

Charubel & Sepharial
Degrees of the Zodiac Symbolized

H.L. Cornell, M.D.
Encyclopaedia of Medical Astrology

Nicholas Culpeper
Astrological Judgement of Diseases from the Decumbiture of the Sick, *and,*
Urinalia

Dorotheus of Sidon
Carmen Astrologicum, *translated by David Pingree*

Nicholas deVore
Encyclopedia of Astrology

Firmicus Maternus
Ancient Astrology Theory & Practice: Matheseos Libri VIII, *translated by*
Jean Rhys Bram

William Lilly
Christian Astrology, books 1 & 2
Christian Astrology, book 3

Claudius Ptolemy
Tetrabiblos, *translated by J.M. Ashmand*

Vivian Robson
Astrology and Sex
Electional Astrology
Fixed Stars & Constellations in Astrology

Richard Saunders
The Astrological Judgement and Practice of Physick

H.S. Green, Raphael & C.E.O. Carter
Mundane Astrology: *3 Books*

If not available from your local bookseller, order directly from:

The Astrology Center of America
207 Victory Lane
Bel Air, MD 21014

on the web at:
http://www.astroamerica.com

Printed in the United States
141073LV00005B/7/A